INDUSTRIAL AND SPECIALTY PAPERS

Volume III—Applications

Prepared by a Staff of Specialists
Under the Editorship of

Robert H. Mosher

Director, New Business Development
Kimberly-Clark Corporation

and

Dale S. Davis

Professor Emeritus, Pulp and Paper Technology
University of Alabama

CHEMICAL PUBLISHING COMPANY, INC.
New York   1969

Industrial and Specialty Papers, Volume 3, Applications

ISBN: 978-0-8206-0168-7

Chemical Publishing Company:
www.chemical-publishing.com
www.chemicalpublishing.net

First edition:
© **Chemical Publishing Company, Inc.** – New York, 1970
Second Impression:
Chemical Publishing Company, Inc. - 2013

Printed in the United States of America

CONTRIBUTING AUTHORS

Bror E. Anderson, Vice President and Technical Director, Weber Marking Systems, Inc., Arlington Heights, Ill.

Glen Augspurger, Manager, Product Development, Champion Papers, Knightsbridge, Hamilton, Ohio.

Rufus Blount, The Hampton Glazed Paper and Card Co., Holyoke, Mass.

Russell S. Bracewell, Vice President, The Marvellum Co.

Gerald Cole, Former Consultant, Highland Park, Ill.

John H. Davies, Manager, Laminated Foils, Revere Copper and Brass, Inc., Brooklyn, N. Y.

Fred W. Farrell, Vice President, McLaurin-Jones Company, Brookfield, Mass.

Francis C. Heywood, President, The Marvellum Company, Holyoke, Mass.

D. R. Hurley, Vice President, Manufacturing, Interchemical Corporation, Cincinnati, Ohio.

D. F. Kramer, Technical Department, Thilmany Pulp & Paper Company, Kaukauna, Wis.

M. A. Kuclar, Technical Director, The Birge Company, Inc., Buffalo, N. Y.

Hugh E. Lockhart, Assistant Professor, School of Packaging, Michigan State University, Lansing, Mich.

William Louden, Vice President, Research, Paterson Parchment Paper Company, Bristol, Pa.

Robert H. Mosher, Director, New Business Development, Kimberly-Clark Corporation, Neenah, Wis.

David S. Most, Consultant in Reprography, Marblehead, Mass.

John C. Pickering, Jr., Vice President, The Pervel Corporation.

R. Heath Reeves, Thilmany Pulp & Paper Company, Kaukauna, Wis.

George Schmidt, Chemist, The Cellutin Corporation, Holyoke, Mass.

Robert H. Simmons, Silver Spring, Md.

Arthur M. Worthington, formerly of American Tissue Mills.

Foreword

Paper is the generic term for a felted sheet of plant fibers in which its thickness is invariably minute compared with its width and length. Where the sheet has been altered significantly by the addition of nonfibrous materials, either by admixture into the felted fibers or by application of coatings or films to the fibrous web, the product is considered a specialty paper.

This volume of *Industrial and Specialty Papers* is concerned with characterizing the products of this sector of the paper industry—defining what they are, how they are used, and where they are sold. In this part of the industry the distinctive products are many and the basic general information is voluminous, but the "know-how" on specific sheets is often the proprietary information of individual manufacturers. For this reason the experienced personal touch of the editors and authors was necessary to render the material into an integrated, useful compilation of information about specialty papers.

As the needs of our society become more complex, the varieties of such papers become correspondingly more numerous and more sophisticated. The growth of these papers in the fields of communication, reprography, graphic arts, packaging and functional industrial applications attests to their importance in filling real needs.

The field of specialty paper uses many products of the chemical industry. With each new and improved chemical product, new opportunities occur. Since the publication of Robert Mosher's *Specialty Papers* in 1950, the first book of its kind in the English language, advance in all phases of the industry has been so consistent that updating with the present volume is necessary to meet the needs of a wide audience.

The very nature of the specialty paper field is such that no one author would feel qualified to write the entire text on the subject. Editors Robert H. Mosher and Dale S. Davis have wisely selected a group of twenty-one experts to present the available information in a form invaluable to the business man, the technologist, the marketing man, the teacher, and the student. This book is essential reading for anyone presuming to be informed about specialty papers.

Harris O. Ware
Executive Vice President,
Premoid Corporation,
Fellow, Technical Association
of the Pulp and Paper Industry

Contents

chapter 1

Printing Papers

R. H. SIMMONS

Printing is the act of reproducing a design on any surface. This chapter endeavors to relate printing papers to the method of printing, the press, and the ink.

The five basic methods of multiple reproduction or printing are: **(1) silk screen** or **screen printing,** where a stencil is mounted on a screen material and the ink is squeegeed through to make a print; **(2) electrostatic printing,** where the image is electrostatically transferred to the paper; **(3) typographic** or **relief printing,** where the printing area is raised above the nonprinting area; **(4) planographic printing,** where the printing and nonprinting areas are in the same plane; **(5) intaglio printing,** where the printing area is below the nonprinting area.

The relation of the ink and paper to the press for each method of printing is discussed. Because large-scale production of printing is done by the first three methods, only short resumes of silk screen and electrostatic printing are given here. The remainder of the chapter deals with typographic, planographic, and intaglio printing.

SILK SCREEN PRINTING

Silk screen printing has advanced from hand production to fully automatic printing. Screen work can be reproduced on almost any material if the inks are formulated to meet the requirements. The inks are "short" so that they print without running and squeegee with little resistance. A thick film of ink is applied to the image area. The thickness of the ink film is determined by the thickness of the screen material and the mesh size.

ELECTROSTATIC AND ELECTROPHOTOGRAPHIC PRINTING

Electrostatic and electrophotographic printing[1] find considerable use in copying, the preparation of offset masters recording by cathode ray tube, and enlarging from microfilm. Recent developments in mechanical design have brought electrostatic printing into commercial work. The U.S. Army is operating an electrostatic printing press to print multicolor maps.[2] Printers of packaging materials are also adapting the method to printing on irregular shapes.[3]

The typewriter is used to produce masters for duplicating by means of gelatine, liquids, stencils, and offset.

TYPOGRAPHIC OR LETTERPRESS PRINTING

Typographic or letterpress printing is done from relief or raised

Fig. 1-1.   Miehle V-50X extra vertical letterpress (*Courtesy of Miehle-Goss-Dexter, Inc.*).

Fig. 1-2. Cottrell double-ender letterpress (*Courtesy of The Cottrell Co., Division of Harris-Intertype Corp.*).

printing areas. (Fig. 1-1 and 1-2). It is the oldest form of printing, dating from 1450 when Johann Gutenberg invented movable type. In early print shops paper was in short supply; handmade rag papers with a laid watermark were all that could be obtained, yet the printer with his single-impression hand press was able to produce an acceptable piece of printing for those days.

As the speed of the presses and the quality of printing advanced, so did the need for better printing papers. They must show increased uniformity in formation, thickness, finish, and receptivity to ink. The invention of the halftone screen gave the printer a tool for reproducing illustrations with enhanced fidelity. To obtain the best results smooth papers are necessary. Perfected process color printing calls for the best uniform paper.

Paper finishes have progressed from antique to machine-finish, English-finish, and supercalendered to coated. Halftone screen rulings are coarse or fine line depending upon the finish of the paper. Newsprint takes from 50-line to 85-line screens; bond, ledger and writing papers, 85-line to 100-line screens; machine-finish and supercalender paper, 100-line to 133-line screens; and coated paper, 120-line to 150-line screens.[4] Highly developed blade coaters now yield smooth finishes with decreased coating material. Research to develop better coated papers is still going on.

Most grades of paper can be printed on typographic presses, which handle roughly half of the printing done today. The presses vary from the hand-fed platen press to the automatic, flat-bed cylinder, and rotary sheet-fed to the high-speed newspaper and multicolor presses. The work can vary from straight type to four- and five-color process printing.

Paper

For all sheet-fed presses the paper must lie flat and be free from wrinkles, tight and loose edges, curl, and extraneous material such as cutter dust and trim. The body of the sheet must withstand the tensions developed in the press. Weak, floppy sheets and curled paper cause hangups, misregister, and even ball ups that damage make-ready and plates. Abnormal curl interferes with proper jogging of the paper in the delivery of the press.

Roll paper for web presses must be uniformly wound with clean-cut edges, free from slitter dust, slivers, and trim. The paper must have uniform formation, thickness, and finish. Variations in thickness produce soft or spongy areas; loosely wound rolls create uneven tension on unwinding. If the roll is loosely wound, telescoping occurs and a movement of the roll to the right or left brings the edges of the web into contact with the side of the press and out of register at the impression cylinder. All splices must be well made and properly flagged.

Rolls with tight or loose edges tend to wrinkle and crease when run through the press. Tight edges can cause the sheet to tear because of the additional tension on the edge. If the edges are tight the middle of the web is longer and sags or bulges. Loose edges are longer than paper in the middle of the web and wave or flop during printing. Both tight and loose edges are the result of uneven tension across the web. If either condition is severe, registration is affected and uneven cutoff at the delivery end of the press occurs.[5]

Inks

Printing inks used on typographic presses vary with the speed of the press, the type of paper, and the requirements of the form being printed. Job press inks should possess as heavy a body as the paper will permit without "picking" or pulling the sheets out of the grippers. The ink must be "short" and buttery with just enough "length" to permit good distribution. As the speed of the press increases, the inks become more fluid with less tack.

Web press inks must be quite fluid consistent with the press speed and the surface strength of the paper. As the speed of the press increases, the uniformity of the paper must improve to give better receptivity to ink. Typographic inks are manufactured so that they print satisfactorily with the various papers required in the finished work.

Typographic inks are classified according to the type of press and the grade of paper used as well as the form that is reproduced. Thus inks are made to operate on job, flat-bed, cylinder, rotary and web presses; on bond, ledger, coated, supercalendered and parchment papers; and on carton stocks. Forms are type, halftone, or solid.

PLANOGRAPHIC PRINTING

Planographic printing is done from a flat surface, i.e. the printing and the nonprinting areas are in the same plane. This process was originally called lithography because the image was drawn with a greasy ink on a smooth block of limestone.

Fig. 1-3. Miehle 49 four-color offset/letterset press (*Courtesy of Miehle-Goss-Dexter, Inc.*).

Lithography

The first printing was directly from the stone to the paper. Then the offset process was developed, in which the printing plate prints on a rubber blanket, which then transfers or offsets the ink to the paper under pressure of the impression cylinder. The cumbersome stone is thus replaced by a thin metal plate, but the process is still called lithography or offset lithography. The inks are lithographic or offset and the paper is specifically for the offset process.

When a lithographic plate is developed, the image or printing area accepts ink and the nonprinting area accepts water. The oil and water immisciblity is the basis of the lithographic process. A plate, a blanket, and an impression cylinder make up an offset press. (Fig. 1-3 and 1-4). The plate is locked into position on the plate cylinder and is alternately dampened with fountain solution and printing ink.

Fig. 1-4. Miehle 49 four-color offset/letterset press

(Courtesy of Miehle-Goss-Dexter, Inc.).

The plate cylinder is run in contact with the blanket cylinder until the press is properly inked up. The paper is then fed between the blanket and impression cylinders where the ink is transferred or off-set from the rubber blanket to the paper.

The lithographic process is a physical-chemical method of printing. A thin film of ink and a thin film of water are in constant contact. Mixing of the two ingredients, or emulsification, can take place. Factors that affect the quality of printing are bleeding of the ink, washing out of the pigment, sensitizing of the nonprinting areas, and spreading or growing of the image.

Paper

Paper for lithography must lie flat and have good dimensional stability.[6] Paper with curl, loose edges or tight edges tends to wrinkle when passing between the smooth surfaces of the blanket and impression cylinders. Paper with loose edges tends to fan or stretch at the back of the roll. The paper must have a certain amount of water resistance to withstand the effect of the fountain solution. Uncoated papers must be highly sized and coated papers require a water-resistant adhesive to bind the coating. The paper must resist the higher tack of litho inks and the tack developed in the rubber blanket.

Coated papers show a greater tendency to pick when dampened with water than do uncoated papers. Wetting agents are added to some papers to permit better wetting of fillers or coating pigments and thus help distribution in the paper. Small amounts of wetting agents in the paper can break the interfacial tension between the ink and water and can cause emulsification of the ink.

Sheet paper for offset must be trimmed squarely on four sides to permit registration on single impression, two-sided work. The paper must not contain loose particles or dust and must not pick or split. Hickies, small white doughnut-shaped spots with inked centers, can be caused by loose particles, pick, or ink skin. Loose fibers from uncoated papers can attach themselves to the rubber blanket. Because they absorb water, they produce white spots without the inked center.

Ink

The very nature of the printing operation requires highly concentrated ink so that the film of ink carried on the plate can be kept as thin as possible to prevent squashing or slurring under pressure. Because of this concentration of pigment or color, "longer," more viscous vehicles must be used. Some water-resistant varnish is usually incorporated in offset inks to help resist emulsification.[7] Thin inks can "grease" or "scum" and cause difficulty in keeping the work open so that the image tends to thicken and grow. If the ink is not water-resistant the acid of the fountain solution undercuts the image and makes it thin or sharp. The pH and amount of fountain solution to reach the printing plate affect the quality and number of impressions obtained.[8] The cost of preparation of the cylinder and the speed of the press make this process feasible only for long runs.

INTAGLIO PRINTING

Intaglio printing is done from recessed plates and rolls. The printing area is cut or etched below the surface of the plate or roll.

In printing from an engraved plate, it must be rolled up with a heavy film of intaglio or plate ink so that the ink completely fills the engraving. The excess ink is wiped from the surface of the plate, which is then polished to remove any ink that remains on the nonprinting areas. The sheet of paper to be printed is saturated with water and placed over the inked plate, which is then run under the impression roll where the paper is pressed into the engraving in close contact with the ink. The ink adheres to the paper and is pulled out of the engraving, producing the characteristic engraved impression. The ink lies on the surface of the paper and is several thousandths of an inch thick, depending on the depth of the engraving. This method is used for copper or steel plate engraving, largely done by hand.

Die stamping or intaglio printing is divided into two main classes, plate printing and gravure printing, and is done by hand- or power-operated presses. The paper is registered by hand and removed by hand. The ink is quick drying and similar to a heat-set ink. Wedding announcements, letterheads, calling cards, and greeting cards are the normal field for die stamping presses, but special work such as multicolor seals, citations, and diplomas is also done on them.

Plate Printing

Plate printing has been mechanized and mass production is being done on single-color, sheet-fed, and web-fed rotary presses. Many multicolor plate printing presses are in use.

Paper

Plate printing is usually done on high-grade, 100% rag papers, such

Fig. 1-5. Miehle ten-color light materials, roll-to-roll rotogravure press (*Courtesy of Miehle-Goss-Dexter, Inc.*).

as bond, ledger, and bristols, but others can be used. Securities, currency, and stamps are mass-produced by means of this form of printing.

Ink

Inks for plate printing are "short" and buttery with a maximum concentration of pigment. The harder pigments and extenders are needed in plate printing to assist in wiping and polishing the plates. On web-fed rotary presses, heat-set inks are used.

Gravure Printing

Gravure or rotogravure printing is an intaglio process in that the printing area is etched below the surface of the printing roller on copper cylinders in the forms of square or round cells. (Fig. 1-5 and 1-6). The depth of the cell determines the amount of ink carried and

Fig. 1-6. Cottrell Model V-22 newspaper press (*Courtesy of The Cottrell Co., Division of Harris-Intertype Corp.*).

the depth of the color printed. Most gravure printing is done on web presses, but some sheet-fed gravure presses are used. The cost of preparation of the cylinder and the speed of the press make this process feasible only for long runs.

Paper

The paper must be uniform in formation, thickness, and finish. It must be smooth and the density of the paper must insure good ink receptivity and must not permit the ink to spread or feather. Sunday supplements to newspapers use rotonews or English finish newsprint. Supercalendered newsprint and supercalendered book papers together with coated papers made especially for rotogravure printing are employed for catalogs and magazines. Gravure is being used for printing paperboard for packaging. Because gravure printing is primarily a web-press operation, C. A. Thompson[9] lists some requirements for gravure newspaper rolls and paper:

1. The rolls should be free from defects such as slug holes, calender cuts, calender scabs, edge cracks, turnover, defective mill splices, winder snapoffs, and loose starts at the cores.

2. The paper should have sufficient strength to withstand the tensions and the stresses that are encountered on a rotogravure press.

3. The rolls should be as uniform as possible and should run with no evidence of slack areas, and baggy or loose ends, which result in wrinkles, misregister, or web breaks.

Gravure Inking

Gravure inks are composed of natural or synthetic resins and gums dissolved in volatile solvents together with the proper coloring materials. The inks are thin, almost without tack, and entirely free from gritty or hard particles, which could scratch the cylinder.

Printing is done directly from the cylinder. The paper passes between the etched cylinder and an impression roller, which presses the paper into the engraving. The impression roll is covered with a slightly compressible covering that helps to force the paper into the cells and compensates for slight variations in its thickness. The cylinder is inked by rotating directly in the ink fountain or by some other method that completely fills the cells of the cylinder. The excess ink is removed by a doctor blade to wipe the nonprinting area clean. The ink is removed from the cells by capillary action, pressure, and affinity of the ink for the paper surface.

Tensions

Tensions in gravure presses are greater than those in typographic web presses. Tension, greater between impressions and about three times as great as at the web stand, increases materially just before a break. Because of the economics of gravure printing and the high cost of preparation, the presses are run twenty-four hours a day. A web break takes about half an hour to repair. Nonuniformity across the sheet with baggy paper creating wrinkles, and poorly-made mill splices accounts for a majority of web breaks on gravure presses. A study of the mechanics[10] of the process has shown that the effects of the nip give rise to shear as well as tension stresses in the nip.[11]

Paper Characteristics in Relation to Printing

Consideration is given here to characteristics that must be built into the paper for satisfactory operation because of the demands imposed by the various printing processes.

Physical Strength

Strength is important because the paper must withstand the stress and strain of the printing and binding operations. Strength is not as important in printing as it may be in the ultimate use of the finished product. In building a paper for printing, one must consider the use requirements of the printed article and then consider these in relation to obtaining the printing properties. Maximum strength and a closed formation with a uniform, smooth finish are impossible. Paper to be folded, stitched, or sewed must have sufficient strength to withstand these operations without breaking.

Formation

Formation is basic for printing papers because many printing characteristics depend on it. Formation is the measure of uniformity of distribution of the components of a sheet of paper, more especially the fibrous part. Formation is normally judged when observing the sheet by transmitted light. A cloudy appearance from irregular distribution of fibers is called a "wild" formation. A uniform appearance is called a "close" or "closed" formation and is highly desirable in printing papers. Because printing presses are designed to give a uniform impression across the printing form, a paper uniform in finish, formation, and thickness is necessary to take a uniform impression. A sheet with a closed formation usually has these characteristics, supercalenders uniformly and accepts a smooth coating.

A "wild" formation is the result of long fibers lumping together to produce high and low spots in the papper. If the paper is calendered, the high spots take the pressure of the rolls and develop a polish while the rest of the sheet remains dull. If too much pressure is applied or too much moisture is in the paper, it tends to become somewhat transparent and darkens in color. The sheet becomes crushed and shiny at the high spots. These spots do not accept printing ink readily and it does not penetrate into the paper. The ink dries slowly with a complete gloss holdout and may offset in printing.

The low or dull spots are not calendered as hard as they should be and remain dull areas between the shiny spots. They accept ink readily and absorb it into the paper. The ink sets readily and will not offset, but because of the penetration it appears flat and dull. The combination of penetration on the one hand and transparency on the other may be responsible for showthrough of the printing. A solid printed on this paper dries with a mottled finish of glossy spots and flat spots.

An uncoated paper can be two-sided in color, finish, and ink-receptivity. The wire side is usually slightly rougher and more open than the felt side. The felt side has always been considered the better printing side of paper; manufacturers of printing papers label and pack sheet paper with the felt side up. Decided differences in the two-sidedness of paper may make it necessary to adjust the ink formula so that the printing on the wire side matches that on the felt side in color or finish.

Electronic scanning of a sheet of paper and charting the results on a recorder are the most successful means of determining and recording the formation of paper.[12] Visual comparison of the paper with standard samples is the accepted method for rating formation.

Smoothness

As stated under *Formation*, p.13, smoothness is closely related to the uniformity of the formation and is relative. One sheet is said to be smoother than another when more points on the surface of the first sheet are more nearly in one plane than are those on the surface of the second. The smoother the sheet the better is the contact between the printing plate and the paper.[13]

Levelness

Levelness, closely allied with smoothness, is variation in thickness from one spot to the next, whereas smoothness is overall variation of small surface imperfections.

The smoothness of coated paper is essential for faithful reproduction of four- and five-color process printing, and is measured by four methods:[14]

Basis	Instrument
1. Air leakage under a smooth ring	Bekk, Bendsten, Gurley, Sheffield, Brecht[15]
2. Stylus tracing for contour	Brush, Talysurf, Metrsurf, Proficorder
3. Optical contact with glass	Chapman
4. Low-angle shadowing	Sheid

Tests with Methods 1, 2, and 4 are not made under printing pressure. Method 3 ignores low spots. The most satisfactory method of determining printing smoothness is to use the Vandercook No. 4 proof press with controlled thickness of ink film, which is less than normal ink coverage. A print made in this manner shows low spots, uneven thickness, and formation.

Compressibility (Printing Softness)

Softness in paper represents the ability of the sheet to compress and conform to the shape of the printing surface under printing pressure. A soft sheet conforms more easily and takes a better impression than does a hard sheet. The soft paper does not require as much pressure to print as does a hard sheet.

To determine this factor the thickness of the paper is measured under a light load and again under a known pressure. The decrease in thickness is the compressibility. When the thickness is again measured after removing the pressure, the return to the original thickness is called recovery or resiliency. The C & R thickness, compression and recovery tester[16] gives an indication of softness under low pressures. The Armstrong indentation machine[16] uses pressures nearer those of printing.

The Gurley-Hill S.P.S. tester,[16] based on the air-flow principle, has an attachment that measures the softness of paper. A specimen is clamped between two smooth circular rings and four steel pins that project 0.002 in. above the surface are set in the lower ring. With no paper between the rings, air flows at a standard rate. With paper between the rings, any penetration of the pins restricts the flow of air —the higher the rate, the softer the paper.

Porosity (Air Permeability)[17]

The rate of flow through a given area of paper is a measure of porosity. Formerly this rate was believed to indicate how fast air could leak through a sheet of paper under printing pressure. Paper of high porosity and a certain amount of tooth or rough finish was supposed to print better by the offset process than would paper of low porosity because porous paper would let the trapped air out from under the rubber blanket. Printing on metal foils and coated papers has disproved these reasons for high porosity. For uncoated papers the porosity is an indication of the capillary action of the pores and fibers in the paper and their relation to the absorption of the printing inks. Coated papers are not porous, but absorption of ink by the coating is high because of the capillaries formed between the particles of pigment.

The Gurley Densometer and the Gurley-Hill S.P.S. tester have been adopted for measurements of porosity.[15]

Opacity

Opacity is the property of a sheet of paper that prevents dark objects on or in contact with the back side from being seen from the front. This factor is important where both sides of the sheet are printed as in books and magazines. Good opacity is essential in publications containing advertising matter. The use of mineral fillers improves opacity, smoothness, and printing quality of printing papers but reduces the strength of the paper.

The Bausch and Lomb Opacimeter is the standard instrument for measurement of opacity by the contrast ratio method.[18]

Grit

In typographic printing, abrasive particles in the paper wear the plates and produce scratches in halftones and solids. Ink does not reach the bottom of the scratches, so they appear as white lines. Scratches in gravure plates fill with ink and print. In offset printing the grit may be picked up by the offset blanket and work back to the printing plate where scratches also take ink and print. Some coated papers are not water resistant, hence the moisture in the fountain solution in offset printing softens the coating. Particles of coating are then picked up by the blanket where they accumulate and gradually wear or sensitize the printing plate so that it takes ink in the nonprinting areas. The pigments used as fillers in uncoated papers are also applicable for use in coatings.

Some pigments are harder and more crystalline than others and cause printing plates to wear. Because of the concentration of pigment on the surface of coated paper, it produces more plate wear than does uncoated paper. Coated paper of uniform thickness and finish requires little pressure to transfer the ink from the plate to the paper, thus favoring reduced abrasion. Actual wear tests of pigments can only be run on the press; however, comparative tests are made by means of the Valley wear tester.[19] Grit in paper can be demonstrated by drawing a strip of paper of definite length between two glass slides under a definite pressure. Scratches on the glass indicate grit.

Surface Strength

With uncoated papers the surface strength depends upon the mechanical treatment of the pulp and the degree of sizing, both internal and surface. How well the fibers are bonded in the sheet determines the amount of fuzz, lint, and pick found during printing.[20]

Coated papers can show several types of pick: (1) coating or surface pick where loose powder or chips are lifted by the ink or blanket, (2) breaking of the coating from the base or blisters in the surface of the coating, (3) breaking or splitting of the base paper, and (4) wet pick or piling of the coating.[21]

The IGT,[21] Hercules, Waldron, FOGRA and GATF (LTF) testers all use tack-graded inks to rate the pick resistance of papers. The classical method employs the Dennison and K & N waxes. The LTF tester and waxes do not define pick in relation to the grain of the paper. Paper has a lower pick when the direction of pull is across the grain. Most printing papers in sheet form are cut with the grain lengthwise on the sheet; thus the pull of the ink is in the weaker direction of the paper.[21] The GATF tester has been adapted to give a wet pick test.[22] Print testers relate pick more nearly to press speeds. The higher the speed of the press the lower must be the tack of the ink to prevent picking.

When a proof press is used to determine pick resistance of paper, the ink can be tested with the paper to be used in the printing work. If picking occurs the ink can be modified. Certain test proof plates indicative of the various forms of plates encountered in printing, such as type, halftones, and solids, can be obtained.

Receptivity to Ink

In printing, the receptivity of the surface of the paper to ink is important. Because of the large variety of printing inks manufactured and the variety of finishes and papers used in printing, receptivity to ink varies and can be expressed only as the relationship between a particular paper and a particular ink. Paper that accepts the ink uniformly with light impression has good receptivity to ink and does not need high pressure to force the ink to penetrate the surface.

The covering power of ink is closely related to the ink receptivity of the paper. Covering power is expressed as the number of square inches of paper that can be covered by a pound of ink. When printing rough-finished, uncoated paper the pressman must get the ink to the bottom of the depressions in the sheet by increasing the pressure or the amount of ink on the plate. If the ink is increased the covering power is reduced.

Paper with poor receptivity to ink can be printed by adjusting the composition of the ink. When ordering inks for a new job, one should give the ink manufacturer a sample of the paper on which the work is to be printed.

Inks for offset printing are stronger in color and tackier than are typographic inks and have more body. These characteristics together with the rubber blanket make offset printing suited to tough papers and those with uneven receptivity to ink, thus enabling one to print fine line screens with great fidelity of reproduction.

Oil absorption methods,[24] such as the drop and flotation, were the first used for measurement of ink receptivity. The Vancometer,[25] in which a thin film of oil is spread over a large area, is a refinement of the oil absorption method. The reflectance as rated by a photoelectric cell is taken at a definite time and is called the printing number.

The use of the proof press or print testers is by far the most satisfactory means of determining the ink receptivity of paper. The proof press with a split fountain permits the paper to be tested with two inks at one time.

Moisture in Paper

Moisture in paper must be controlled to balance the conditions required by the print ng process. Paper gives up or takes on moisture to stay in equilibrium with atmospheric conditions. If the paper is to be printed in an air-conditioned pressroom it should be ordered with a controlled moisture content to balance these conditions. When the pressroom is not air-conditioned the moisture content of the paper should be adjusted to meet natural atmospheric conditions, such as high humidity in summer and low humidity in winter when the heat is turned on.

When the paper takes on moisture the edges lengthen, are wavy in the pile, and are called "loose." When printed, the image across the back edge of the sheet is wider than that across the front edge, a situation called "fanning." In extreme conditions a wrinkle may start at the center of the sheet and extend to the back.

When paper gives up moisture the edges shrink and are called "tight." Sheet paper with tight edges shows a baggy middle. When printed the image across the back is shorter than that across the front and is said to "draw in." In extreme conditions, wrinkles may start just back of the front edge and tend to curve toward the center. The baggy part shows smudges in the solids and elongated halftone dots.

Sheet paper with a low moisture content can absorb enough moisture going through the press to expand or stretch around the cylinder. Under normal conditions this expansion would be in the cross-machine direction, since printing papers are purchased with the machine direction lengthwise on the sheet.

During periods of dry weather the very action of the press creates static in the paper. Static is responsible for double feeding, poor registration at the guides, hung delivery, and poor jogging in the delivery. If static is present, the delivery pile attracts the sheet and draws it down onto the pile with considerable force. The ink on the previous sheet does not have time to set properly and the force of attraction brings the two sheets into such intimate contact that the ink offsets on the back of the top sheet. If the paper contains sufficient moisture, the static charge leaks off without causing difficulty. An increase in humidity or the use of a static eliminator is helpful.

The hanging of sheet paper to attain equilibrium with pressroom conditions is helpful in eliminating tight and loose edges. If the sides of the pile are exposed to banks of infrared lamps just before printing, tight edges relax and loose edges flatten out.

Roll paper gives up or takes on moisture through the ends of the roll. When exposed to high humidities the ends of the roll expand and when the roll is unwound the sides of the web are longer than the middle.

The edges then flap or wave in the press. Long edges tend to print off-register and the cut-off is bad. If the ends of the roll give up moisture the edges become tight, and if the tension is too great the web breaks or tears. Registration and cut-off are then poor in the middle of the sheet because it is baggy.[26]

When four-color printing is done on a single-color press, controlled moisture conditions are essential for good registration. This can be maintained in web offset and heat-set printing if the moisture content of the paper is kept lower than that for sheet-fed presses. When passing through the drying ovens the paper does not change dimensions as much as does a paper with normal moisture content. Papers with high moisture content may blister when exposed to the high temperatures.

Moisture content of paper coming off the paper machine can be controlled to compare favorably with pressroom conditions.[23] If properly packaged to prevent changes in moisture content, the paper should retain its preconditioned state for a considerable time.

The moisture content of paper can be determined by means of oven drying[27] and toluene distillation.[28] The sword hygrometer shows whether the paper is in balance with the outside humidity. The Hart moisture meter must be calibrated for each grade of paper to be tested. It can then be used to determine the moisture content of unknowns in that grade. The Hygro-Champ[29] uses a sensing head to determine equilibrium moisture content and thus indicates whether the paper will give up or take on moisture. Other means of determining moisture content of paper use variable-resistance, capacitance-sensing elements; a recent report[30] covers a moisture gage that employs infrared backscatter.

Dimensional Stability

Dimensional stability is the property of a sheet of paper that relates to the constancy of dimensions in the machine- and cross-machine directions under conditions of varying relative humidity. It also covers dimensional changes that are due to mechanical stresses imposed during the printing operation. Good register between colors can be maintained if the paper has good dimensional stability.

Two instruments to measure dimensional stability are the cabinet recommended by TAPPI Standard T447, Moisture Expansivity of Paper, and the Neenah Expansimeter.

Grain

The grain of paper is the machine direction. In roll paper the grain is always around the roll.

When sheet paper is ordered for printing, the grain is normally specified to lie along the length of the press cylinder because the paper is flexible in the cross-machine direction so that the sheet hugs the cylinder closely. Because expansion in the machine direction is less than that in the cross direction, the sheet resists the rolling pressure. If the paper takes up moisture and expands during the first impression, it will not register properly with the second impression. Because there is no side adjustment on the press and the lesser expansion takes place in this dimension, the pressman can adjust his guides to split the variation in registration and thus obtain reasonably good results. To compensate for the great variation around the cylinder the press packing can be increased. Increased press packing lengthens the circumference of the cylinder and expands the impression so that better registration is obtained.

When paper is folded, the fold is smoother along the grain than it is across the grain. When folded into signatures the thickness of the fold along the grain is less than it is in the cross direction. When signatures are bound in a book the grain of the pages is vertical to impart rigidity when the book is standing on a shelf. If the grain is horizontal in a book the pages curl down and sag.

When used for typing or duplicating, sheet paper of letterhead size should have the grain the long or 11-$\frac{1}{2}$- inch way so that the sheet will leave the platen or cylinder of the machine. If the grain is lengthwise of the platen or cylinder the sheet tends to follow the roll and wrap around it.

Grain[31] can be determined by wetting a small piece of the paper. The curl will be around the grain direction. When a piece of paper is folded in the two principal directions the grain is in the direction of the smoother fold. If a strip of paper for each principal direction of the sheet is cut and each held out horizontally, the strip that shows the lesser droop is in the machine direction.

Color and Brightness

To record the color of white paper, a chart of spectral reflectivity is drawn.[32] Another term connected with white papers is brightness,[33] which is the reflectance in the blue and violet portion of the spectrum.

What the printer needs most in printing papers is uniformity in color. Sheet paper is normally gang-cut, several rolls at a time. Any variation in color from sheet to sheet is repeated regularly throughout the skid or package. Mixed color when bound in a book shows up on the edges as striations. If the base white is not uniform, variations can affect the color of the ink, especially when process colors are used. The use of fluorescent white dyes and optical brighteners is standard procedure in printing papers. These dyes introduce another factor in color matching of paper and inks.

Acidity and Alkalinity

In paper, acidity and alkalinity are measured in the pH or hydrogen ion range.[35] Papers in which permanence is required should have a pH of at least 5.0. Most uncoated papers have a pH of less than 7 and are on the acid side. Alum is added to set the rosin size. The dyes used to tint the white paper also require an acid condition to prevent bleeding. In some papers an alkaline filler is employed to improve receptivity to ink. Such papers are not satisfactory for use in lithography because the acid fountain solution would (1) cause the filler to foam, (2) become neutralized, and (3) would not keep the printing plate clean. An alkaline filler can also sensitize the plate.

Coated papers normally have a pH of 7.5-10.0.[34] Paper with a pH in this range increases the rate at which the ink dries.

Uniformity

In printing papers, the following factors must be kept uniform: basis weight, formation, thickness, finish, receptivity to ink, and color. Printing papers must also be free from holes, specks, spots, and surface blemishes. Compensation can be made for variations from the normal if they are uniform throughout the shipment.

Laminating and Varnishing

Laminating printed work between layers of synthetic plastics and varnishing produce scuff resistance in labels, charts, menus, cover papers, and other items that are handled continually. Laminating and varnishing improve the brilliance of the printing. Papers that are to be varnished must have good receptivity to the varnish, but must

also have good gloss holdout. Paper that permits the varnish to penetrate into the sheet becomes transparent and blotchy. For papers that
are to be laminated or varnished, inks that do not bleed or smear must
be used.

Printing Papers

"Any paper suitable for printing" is an all-inclusive definition, but
when printability is defined as the property of a paper that favors printed matter of good quality, the spread narrows down to papers developed
specifically for printing purposes. Typical are papers used in publications where halftones are reproduced in black and white and by the
four-color process. Papers used for office supplies, forms and duplicating work, labels, and some wrapping papers are considered printing papers. The expansion of package printing into multicolor printing on wrappings and on folding boxboard that is coated on one side
raised these to the class of printing papers.

REFERENCES

1. George E. Mott, "Unconventional Image Forming Systems in the Graphic Arts," *TAGA Proc.* (1965), p. 473-500.
2. "A First Hand Report on the Electrostatic Printing Process," *Lithographers J.* **48**, No. 2:22-7 (May 1963).
3. Robert P. Long, *Package Printing*, Graphic Magazines, Inc., (1964).
4. Forest Rundell, "Fit Your Illustration to the Paper on Which It Prints," *Inland Printer* **126**, No. 2:38-9 (Nov. 1950)
5. William H. Bureau, "Papers for Reproduction and Printing Processes." Graphic Arts Monthly, **34**, No. 9:60-6, 194 (Sept. 1962)
6. Harry R. Baldwin, "Pressroom Problems Resulting from Various Paper Defects and Changing Press Requirement," *TAGA Proc.* (1964), p. 182-191.
7. J. Albrecht and M. Heigl, "Water, Ink and Tack," *TAGA Proc.* (1965), p. 337-48.
8. J. H. Bitter, "Emulsification of Offset Inks," *Intern. Bull. for the Printing & Allied Trades* No. 73:24-9 Disc. 29-30 (Jan. 1956).
9. C. A. Thompson, "Gravure Newspaper Supplement Magazines in the United States," *TAPPI* **48**, No. 1:88A-90A (Jan. 1965).
10. Karl A. Springstein, "New Measurements on Gravure Machines," *TAGA Proc.* (1965), p. 270-9.
11. Harvey F. George, Robert H. Oppenheimer, and John J. Kimball, "Gravure Nip Mechanics." *TAGA Proc.* (1964), p. 151-61.
12. The Thwing-Albert Formation Tester, Thwing-Albert Instrument Co.
13. William C. Walker and Robert F. Carmack, "The Printing Smoothness of Paper," *TAGA Proc.* (1963), p. 235-58.
14. "Smoothness of Printing Paper," *TAPPI* T-479.
15. W. Gurley and L. E. Gurley, "Gurley-Hill S.P.S. Tester," *Paper Trade J.* **106**, No. 14:31 (1938), Institute of Paper Chemistry Instrumentation Studies XXXVI; "The Gurley-Hill S.P.S. Tester," *Paper Trade J.* **110**, No. 23:27-33 (June 6, 1940).
16. Catalog, Custom Scientific Instrument Inc.
17. "Air Resistance of Paper," *TAPPI* T-460.
18. "Opacity of Paper," *TAPPI* T-425.
19. Sandford S. Cole, "The Valley Abrasion Tester for Titanium Dioxide Pigments," *TAPPI* **46**, No. 2:124-6 (Feb. 1963).
20. G. L. Swope, "A Method for Determining Pick Resistance in Offset Sheet Stock," *TAPPI* **45**, No. 5-199-201A (May 1962).
21. "Surface Strength of Paper (IGT Tester)," *TAPPI* T-499.
22. Everett M. Bernstein, "Adaptation of the LTF Pick Tester for Moisture Pick Test," *TAGA Proc.* 1963, p. 195-205.
23. A. P. Reynolds, "Effects of Press Moisture on Lithographic Printing Papers," *TAPPI* **46**, No. 7:152-55A (July 1963).
24. "Printing-Ink Receptivity of Paper (Castor-Oil Test)," *TAPPI* T-462.
25. V. V. Vallandigham, "Forecasting Printability by Oil Absorption Measurements," *Tech. Assoc. Papers* **29**:523-5, disc. 73-74 (June 1946); *Paper Trade J.* **123**,No. 17:39-41 (T. S. 209-211) (Oct. 24, 1946).
26. "Paper Troubles Chart for Pressmen," *Modern Lithography* **28**, No. 4:40-1 (1960).
27. "Moisture in Paper and Paperboard," *TAPPI* T-412.
28. "Moisture in Paper and Paperboard by Toluene Distillation," *TAPPI* T-484.
29. "Precision Measurements and Control of Relative Humidity," Hygrodynamics, Inc.

30. Arthur J. Beutler, "An Infrared Backscatter Moisture Gage," *TAPPI* **48,** No. 9:490-3 (Sept. 1965).
31. "Machine Direction of Paper," *TAPPI* T-409.
32. "Spectral Reflectivity and Color of Paper," *TAPPI* T-442.
33. "Brightness of Paper," *TAPPI* T-452.
34. Robert F. Reed, "How to Lithograph Coated Offset Paper," Publication by Kimberly-Clark Corp. (1954).
35. "Hydrogen Ion Concentration (pH) of Paper Extracts," *TAPPI* T-435.

chapter 2

High-Gloss Specialty Papers

I. Introduction

R. H. MOSHER

Among the early decorative and fancy papers were the high-finish, glossy papers that were used for wrapping and label work. This type of sheet was first produced by means of the flint-glazing process, which later led to the development of friction-glazing and the brushed enamels. These processes were all based upon the technique of coating a base paper with a plain or colored pigmented or dyed coating compound that could be burnished or polished in a later operation so as to produce a smooth glossy surface. The formulation of the coating is important as the final surface effect is directly dependent upon it. The formulas have always been closely protected secrets, and even today each mill carefully controls and guards its own formulations. The mechanical finishing operations all involve a surface polishing technique for burnishing the coated surface, which contains a certain amount of lubricating and gloss producing agents, until the desired finish is obtained. This burnishing operation involves three basic principles with the three different types of paper.

FLINT-GLAZED PAPER

A polished stone is drawn back and forth across the surface of the coated paper. This action burnishes the surface and produces the flint-glazed paper.

FRICTION-GLAZED PAPER

Coated paper is passed through a friction glazing calender. This consists of two or three rolls—a metal and a paper roll or a paper and two metal rolls. The rolls are driven at differential speeds with the metal rolls operating at about one and a half to four times the peripheral speed of the paper rolls. The coated face contacts the metal or polishing roll and the differential speed produces slip and a burnishing effect.

BRUSH-ENAMELED PAPER

Coated paper is passed through brushing machines where the coated surface is polished by means of a series of rotating brushes of varying degrees of stiffness. Then it is supercalendered and a satiny glazed surface is produced.

These processes are all relatively slow and correspondingly expensive. Therefore work has been done on the development of coated papers that can be mechanically treated at relatively high speeds, after coating, to produce a glazed surface. The original offerings were mainly in the field of white papers, but colored sheets have also become widespread. There are still certain limitations as to the types of pigments that can be used as the sheet must stand extensive calendering and in some cases buffing or brushing to develop the required surface finish. The speed at which the paper can be produced is a real improvement over the prior methods, but two operations—a coating and a finishing stage—are still required.

Another major development in the glossy paper field has resulted in the production of a high-finish paper in one operation. This process has been carried out in several different ways, but all are based upon the same principle. The basic technique involves laying down a film of pigment and adhesive in the normal manner and then pressing this wet or partially dried film while still in a plastic stage against a polished mirror-finished surface and completing the drying in this position. The resulting coated paper, when stripped from the polished surface, retains a mirror image of the casting surface. The polished surface may take the form of a rotary drum drier, an endless metal belt, or a precast impervious film or coated paper. Although the rate of production has been generally rather low on most of the equipment developed so far, the process has shown promise. The separate coating and smoothing operations are reduced to a single combined coating and glazing process and the width of the sheet is only limited by the coating and glazing facilities that can be supplied.

Each method of production and the various problems involved will be discussed in detail.

II. Flint-Glazed Papers

R. S. BLOUNT AND B. E. ANDERSON

The flint-glazing process of finishing coated paper so as to produce a sheet with a high-gloss surface finish is one of the oldest known methods of the paper converting industry. The technique originated in Europe about the middle of the 19th century and the major portion of the development work was done in Germany and Belgium. The chief drawbacks of the process are the low-volume output and limited width of the equipment, but no other method has ever been devised that completely replaces this operation. For this reason flint-glazed papers are still being produced in limited volume, even though for white and pastels other methods have been developed that produce high-gloss papers at lower cost.

MECHANICAL EQUIPMENT

The principle involved in producing this finish is one of "ironing" the sheet by means of a polished stone that passes to and fro across the web five to ten times. The friction of this smooth stone passing over the surface under pressure produces the so-called "flint finish" on the coated paper.

The flinting machine (Fig. 2-1) is a small unit about 4 ft square consisting of a flat bed on which is placed a board of hard, resilient fiber. The remainder of the unit consists of an unwind and a rewind stand and tension controls. The precoated paper is passed over the top of the board at a low rate, approximately 7 ft/min.

The stone is mounted in an iron clamp attached to a long pole usually centered on the ceiling of the room. By means of a cam-driven rocker arm, the stone is forced back and forth across the sheet under pressure. The coating must be formulated with the proper amount of wax and talc for the particular pigment used.

The pressure with which the stone is forced down upon the paper surface as it is moved back and forth must be varied with the color and the coating formulation to obtain the best results.

The typical mill for flint glazing has a large number of individual machines because each unit can only produce approximately 3 to 5 reams per shift. This necessitates considerable floor space, and thus by modern methods of streamlined production the flinting technique

Fig. 2-1. Flint-glazing machines (*Courtesy of Wyomissing Paper Co.*).

appears primitive.

BASE PAPER REQUIREMENTS

The base papers best suited to flint glazing are those with a uniform, well-closed formation so that they can accept the coating smoothly. The sheet must be sufficiently sized so that the absorption of the coating is low. Many types and weights of base stock are used including groundwood and kraft, but 45-47 lb (25 × 38—500) book stock is most common. In general, the appearance of the coated sheet is dependent upon the uniformity of coating absorption. Kraft, therefore, gives more mottle.

Groundwood is particularly well suited to flint finishing as the action of the flint stone smooths the surface to such an extent that the rough surface showing outlines of the individual fibers, characteristic of groundwood as a coating base, is not readily seen in a flinted sheet. Groundwood is not widely used, however, except for special applications because of its lower strength.

Most flint coatings are applied by air knife. Formulation is quite complicated. Casein or soy protein are the most common adhesives and all the formulas contain certain amounts of slip agents, generally finely divided talc or various waxes. Based on 100 parts of dry pigment, 4 to 11 parts of talc or 1-$\frac{1}{2}$ to 5 parts of wax (either carnauba or a mixture of carnauba with other vegetable waxes and petroleum wax) in emulsion form are used.

Medium and deep shades are obtained by the use of pulp or lake colors consisting of dyes or pigments precipitated on a flintable base. The latter is formed by coprecipitation of aluminum hydrate and blanc fixe ($BaSO_4$). The ratio of these two whites in the base varies somewhat, depending on the particular dyes or pigments used. Pastel colors may be achieved by either dyes or pigments; the most brilliant flint colors can be achieved only with certain dyes, at the expense of fastness to light. Although flint-glazed paper may be of any light fastness, the largest selling varieties do not fade in 24 hours in a Fade-Ometer (except reds, which start fading in 3 to 4 hours). The most brilliant light reds may show fade at 30 minutes in a Fade-Ometer. Pigments used must be fairly soft and must not be too abrasive or they would require more frequent regrinding of the flint stone.

APPLICATIONS

Flint-glazed papers possess characteristics that make them attractive to both printers and box makers. When this method of finishing is used, the bulk of the paper is retained to a much greater degree than when a similar finish is achieved by the use of the heavy pressures exerted by calender rolls. Thus many of the desirable characteristics necessary for fabricating boxes, labels, box coverings, and packaging are retained in the sheet so that it lies flat and works easily without curling. However, the sheet, particularly in the deep blues and greens, often shows a characteristic mottled appearance.

The question of how much longer it will be economical to produce flint coated paper is raised each time an article is written on the subject. This process may be used for a great many years, but the number of producers and the number of colors stocked may continue to decrease. (The largest producers still stock from 50 to more than 100 colors.) New developments in catalyzed high-gloss lacquers applied by gravure are a threat to large volume flint business, particularly where the paper is printed on the first stations of a gravure press, overlacquered on the last station, and cured in an extended or high velocity drier. Flints and frictions will continue to dominate the short-run, high-gloss, medium-and dark-colored box wraps for the foreseeable future. Methods of producing medium high gloss by direct machine operations in connection with a coating unit or a high-gloss calender are gradually replacing flint in some areas of the set-up box wrap field. One flint producer has switched to the use of a stone roll in a process that resembles friction calendering.

III. Friction-Glazed Papers

R. H. MOSHER AND B. E. ANDERSON

The development of the various glazing processes parallels in many respects the history of the coated-paper industry. In the early days, paper was handcoated in the sheet form and at that time both sheet flint-glazing and sheet friction-glazing were in use as finishing techniques. About 1830, the first continuous friction-glazing calendering operation was reported; since then there has been a steady volume of glazed paper produced by this process.

When friction-glazed papers were first studied on an experimental basis, it was found that the heavy metal and hard paper rolls originally utilized tended to produce wrinkles and hard spots in the papers. Experiments proved that a softer backing roll, preferably composed of a combination of woolen and cotton fibers, did a satisfactory job and most of the early units contained such rolls. In the early work on these papers, the wax was applied to the surface after the coating operation had been completed, either by wax coating or by dusting it on the sheet in a dry form ahead of the frictioning roll. In later years, the wax was incorporated into the coating color and the entire coating applied on the coating machine.

The friction calender (Fig. 2-2) usually consists of three rolls. The polishing or frictioning roll is mounted at the top of the stack and is driven directly from the power source. The bottom roll is geared to the top roll so that the top roll has a speed one and a half to four times as fast as that of the bottom roll. The bottom roll is made of cast iron and is used both to drive and support the paper roll. The paper or backing roll, which usually has a diameter about three times that of the polishing roll, is not directly driven but obtains its motion by contact with the bottom roll. In some of the early two-roll setups, the paper roll was directly geared to the polishing roll and a special heavy steel core

Fig. 2-2. Friction-glazing calender (*Courtesy of Wyomissing Paper Co.*).

was used to support the paper roll. The production rate varies from 200 to 500 ft/min. depending upon the quality and finish desired.

The polishing or friction roll is normally made of cast and chilled iron as highly polished steel and chrome-plated rolls have been found to be relatively unsatisfactory. Apparently, the minutely rough surface of the chilled iron is essential for the production of high gloss. This roll is usually cored so that cooling water or steam can be circulated through it as desired. Pressure can be applied to the roll journals by either mechanical or hydraulic means. Many calenders are fitted with steam jets because humidification of the sheet is often found to be beneficial in the frictioning operation; steam is also used for correction of cure.

Considerable has been done in studying the relationships between the roll pressure and the frictioning ratio to attain the best conditions for glazing. Most of this work has been done in the mills themselves and little published data are available. The general experience is that good glazed paper cannot be obtained without the use of pressure, but that the minimum pressure consistent with satisfactory glazing should be provided. In general, the lower the applied pressure the higher the required friction ratio to obtain equivalent gloss. The lower pressure results in less curl in the sheet and a more uniform appearance.

Different colors and papers have also been found to require variations in machine conditions to obtain the best glazing results. Metallic paper, such as "Argentine silver," usually requires the highest pressures and friction ratios, whereas deep shades require intermediate conditions. Light colors require the least severe conditions to obtain a satisfactory product. In mills where a wide variety of frictioning conditions have to be met, the present procedure involves the use of chain drives or direct drive gears. To obtain the desired finished product, avoid varying the machine conditions other than pressure. Modern practice involves the use of machines with fixed gear ratios and varying the coating color formulation so as to obtain any desired differences.

Almost any desired color shade can be produced by means of this process, and both light-weight papers and boards have been glazed. In general, the extensive working of the sheet between the paper and metal rolls tends to cause curling. If high roll pressures are used, the sheet structure may be crushed to the extent that any nonuniformity of density and formation may be emphasized.

Although cast-coated papers and high-gloss calendered papers have cut into the volume of friction-glazed papers, they are still produced for special purposes. Their particular field of application is in the deep shades and special shades because that type of sheet is more difficult to produce economically on the larger and higher-speed units.

Base papers for friction glazing are similar to those used for the flint process, although sometimes the base paper for frictioning is a little more critical than that for flint glazing. On the other hand, the friction process has also been used for board, whereas the flint process is limited to papers.

COATING FORMULATION AND TECHNOLOGY

Below are given the major requirements of a coating color for use in friction glazing.

Fine Pigment Particle Size

The pigment agglomerates are broken down to approach their ultimate particle size by milling the color in equipment such as a pebble mill or a Kady mill. In general, the better the degree of dispersion of the pigment the higher the gloss that can be obtained.

High-Finishing Pigments

The pigments used in friction calendering formulations are usually relatively soft, finely dispersed types where colored pigments are needed. The high-finishing clays, satin white, high-finishing calcium carbonates, and titanium pigments are used where white pigments are desired.

Adhesives

The adhesives used should possess high brightness and high bonding strength, because a minimum amount of adhesive (7 to 12%) is used in the formulation to allow glazing of the pigments and maintain brightness. Binder ratios for friction are appreciably less than for flint for the same wax pick.

Slip Agents

High-finishing waxes, such as beeswax or carnauba wax, are included in the formulation to assist with the polishing operation. Talc is also used. About 2 to 5% of talc and between 5 and 9% of wax based upon the dry pigment are commonly used.

The following are fairly typical formulations used in friction finishing.

1. *White Friction Finish*
    English Clay Slurry (60% solids)   170 lb
    Talc   3 lb
    Red Dyestuff (10% solution)   2 qt
    Beeswax Dispersion (12.5%)   5.5 gal
    Casein Solution (18%)   8-12 gal

2. *White Glazed Finish*
   High-Finishing Domestic Clay   200 lb
   Titanium Pigment   32 lb
   Carbonate Pigment   150 lb
   Water   45 gal
   Sodium Hexametaphosphate   1 lb
   Talc   12 lb
   Beeswax and Carnauba Dispersions (12.5%)   9.5-15 gal
   Casein Solution (18%)   32 gal

3. *Black Friction Finish*
   Black Special Frictioning Pulp (53% solids)   240 lb
   Talc   6.5 lb
   Nigrosine Dye (5% solution)   9 oz
   Water   14 gal
   Carnauba Wax Dispersion (12.5%)   8 gal
   Casein Solution (18%)   10-11 gal

To these formulations must be added the various surface-active
agents, plasticizers, preservatives, and other materials necessary to give
optimum operating conditions on the coater.

Applications

Friction-glazed papers are used in box wrap and tube wrappings,
in applications where some tendency of the paper to curl can be tol-
erated. Although they are always less expensive than comparable
flint papers, they are sold in less volume.

IV. Brush-Enamel Papers

F. C. HEYWOOD

When the flat bed stone lithographic process was popular, surface-
coated papers were a prime necessity, especially for color work. Ordinary lithographic plate paper was generally used for much of the work,
but where fine work in multicolor was required, the regular plate finish
was not high enough to do a satisfactory job and produce the best
result. Brush-enamel papers were developed. These had a higher
gloss and finish and such papers were produced and used extensively
in the second half of the 19th and early part of the 20th centuries.

Like the lithographic process itself, the new development came from
Germany. The art of paper coating had been developed there to a
high degree, and the brush-polishing process, which was first used for
single sheets and later on full webs by the Germans, soon was brought

to this country. The production of these papers which involved coating the sheet with a compound that could be burnished to a high polish and then calendered was carried out in a few of the high-grade mills where they were further developed.

BASE PAPER

The base paper used was a good sulfite-soda sheet. Its formation was uniform and well closed up and the sheet was well-sized to hold the coating. The surface was also smooth and level so as to take the exceptionally smooth and uniform coating.

COATING MATERIALS

The English china clay used as the base of lithographic plate paper was replaced in brush enamels by higher and higher percentages of satin white, a synthetic pigment made by the reaction of alum and lime. English china clay and the aluminum hydroxide acted as a good finishing agent. As the commercial production of satin white became more standardized under better manufacturing control, the coating superintendent was able to use higher percentages of it in his coating formulas. During the first decade of the 20th century, when the manufacture of brush-enamel paper was at its height, some coaters were using practically all satin white and no china clay, with perhaps an additional portion of blanc fixe to give body to the coating.

This was a pigment formulation difficult to use, especially after casein had displaced animal glue as the sizing agent. It was a temperamental composition to handle, and as practically no coating plant of those days had what would be called a chemical staff, its use was mostly a matter of trial and error. Fortunate was the coating superintendent who could run—day after day, batch after batch—a high-percentage satin white formula without getting into trouble, as the mix often thickened up with a false body that could not be thinned by any reasonable amount of extra water.

The satin white itself, though giving good lustre, was light and weak and rather hard and brittle so that some additional filler was necessary to yield the richest, most even surface. Less blanc fixe than china clay was required to do this so that a higher percentage of satin white could be used. As both satin white and blanc fixe were considerably more costly than china clay, the formulas had to be governed by economic factors. If cost was not important, a very good formula for a high-finish, smooth surface called for 80% satin white and 20%

blanc fixe.

More sizing was required to bind this mix than to bind those that depended on 100% clay. A lubricant had to be used in the formulation to give the best results in the brushing process. A very finely divided talc powder to the extent of 5% of the weight of filler along with 1 to 2% wax emulsion and saponified castor oil were added. The talc dusted off during the brushing process, whereas the wax and oil helped as a lubricant as well as a plasticizer for the hard, brittle satin white.

Many problems face the brush finisher in connection with his choice of coating pigments. The pigment particles must not be too hard and sharp or excessive brush wear results. On the other hand, if they are too soft or insufficiently bound, excessive dust is produced, resulting in dusty paper and reduced finish. The amount of adhesive must be adjusted to permit a high supercalendering after the brushing and yet the adhesive must be sufficiently hard to bind the pigments during the brushing operation. Lubricants such as mineral oil emulsions or hydrogenated fish oils are sometimes used to reduce dusting and increase brush life, although this is not a general practice.

The moisture content of the sheet being brushed should be high enough to prevent the formation of excessive static. Steam is sometimes blown on the brushes to help in this respect.

COATING TECHNIQUES

The coating process is the same as for other types of surface coating. At the time brush enamels were at their fullest demand, brush-coating technique was the usual coating method, but there is no reason why the air doctor, roll, or spray coating techniques should not be satisfactory.

The high satin white content of these coatings slows up the drying considerably, so that additional drying capacity is necessary or lower production results as compared with ordinary lithographic plate paper.

BRUSHING AND FINISHING OPERATIONS

The only mechanical equipment needed in addition to the regular coating and finishing equipment of a coating plant is the brush machine, (Fig. 2-3). The first brush machines in use operated on the cylinder or drum principle. Later developments of the brush machine were for flat bed or arch-back machines of considerable length where

Fig. 2-3. Angle view of brushing machine (*Courtesy of John Waldron Corp.*).

more brushes could be used and higher speed could be obtained with
satisfactory burnishing.

A machine a with drum of some 48 to 60 in. diameter was set up with
six or eight cylinder brushes around its periphery so that with an unwind
stand and a reel a very workable brush machine was obtained. The
coated paper is fed around the drum and on the reel. The drum travels
with the paper. The brushes are adjustable so that the pressure they
exert on the paper passing under them can be varied. This pressure
is only sufficient to make the tips of the brushes just whip the paper
surface. The brushes are then driven at high speed in counter direction to the flow of the paper. The brush speed may range from 2000
ft/min peripheral speed to nearly 7000 ft/min on some modern installations. The paper can be run through such a brush machine at
200-500 ft/min and receive a very satisfactory burnishing. One man
and helper can take care of several machines at such speeds.

The brushes are usually graded from a fairly stiff hair bristle for
the first one contacting the paper to a soft badger hair for the last,
although bristle brushes have been used. The bristle material was
usually imported pig or similar bristle. Nylon now replaces the natural
bristle. It wears longer and if the ends of the bristles are rounded it

seems to impart a burnishing action that improves the finish. For different types of pigment, brushes of different face densities and length of bristle are used.

The final finishing is done by supercalendering as in the handling of litho-plate paper. It is a combination of the satin white in the coating, the burnishing by the brush machine, and the supercalendering that produces the glossy lustre of the brush-enamel paper.

With the development of the rotary offset lithographic process, a radical change occurred in printing paper requirements. The offset process required no coated surface to produce its best results. In fact, a surface-coated paper was not desirable to give the soft tones of the offset and it was a detriment in obtaining the speeds at which the offset presses were designed to run.

Therefore, as the stone bed lithographic presses passed into history, the demand for lithographic plates and brush enamels diminished rapidly. Only a limited amount of this paper is now produced, and mainly for other purposes. Eventually its production may become one of the forgotten arts.

V. Cast-Coated Papers

GLEN AUGSPURGER

The basic objective of cast-coated papers is the production of a paper possessing a high-gloss finish for use as a printing paper and as a converting base stock. The process generally involves applying a fluid coating to the paper and then solidifying this coating while it is in contact with a polished, nonadhering surface, so that the finished coated surface is a mirror reproduction of the surface upon which it dried.

The weight of coating applied may range from 1.5 lb per 1000 ft² to 7 lb per 1,000 ft², and the major problem with the heavier coating loads in this process is the reduced rate of drying. In many cases the paper first receives a base coating on a paper machine or off-the-machine coating unit and then the final cast coating is applied. With this type of multiple-coating operation the base coat may be colored by means of color pigments or soluble dyestuffs or an over-all colored printing or coating job can be done before applying the cast coating.

The coated surface must harden n contact with the casting surface. The coating and the surface should be of such a nature that the paper will stick to the solid surface of its own accord until set and then break loose spontaneously. The moisture content of the coating at the point where the sheet is stripped from the casting surface is quite critical and must be controlled within narrow limits for successful operation.

BASE PAPER

The base paper for use in cast coating may be of any desirable basis weight or fiber composition. The physical strength will depend upon the end use of the coated product. Its surface should be as smooth as possible with no surface fuzz, and in general practice the sheet is often machine calendered and even supercalendered before cast coating. It should be sized sufficiently to hold up the coating and should not curl when the base coating is applied. A commonly used base stock is a soda-sulfite sheet well sized with rosin. The surface characteristics have a great deal to do with the amount of coating required to produce a smooth glossy coated finish; the smoother the surface the less coating is required to produce a satisfactory finished sheet.

COATING EQUIPMENT

The coating may be laid down directly on the base stock or on a precoated or printed sheet. The application mechanism may be of the brush, roll, air knife, blade, gravure roll, spray, or other suitable type that will apply a uniform and smooth coating to the surface of the paper. The sheet may be partially dried before it reaches the glazing unit, or the drying may all take place on the casting surface. In some processes the surface of the partially dried coating is rewet just before it is applied to the polished surface.

CASTING EQUIPMENT AND TECHNOLOGY

The actual cast-finishing equipment usually consists of a large polished metal cylinder of the Yankee drier type but may also be an endless metal or rubber belt or the surface of a highly finished impervious sheet. If a metal drum (Fig. 2-4) or endless belt is used, the surface must be ground smooth, polished, and then plated with chromium, nickel, monel, or other metal. This surface is finally polished until flawless. A hard rubber casting belt or apron has also been used. Where an impervious web serves as the casting surface, cellophane, coated paper, cellulose acetate film, or metal foil has been suggested. The method of drying may be internal, as in the metal drum or a heated metal belt, and/or external with heated air or radiant heat directed to

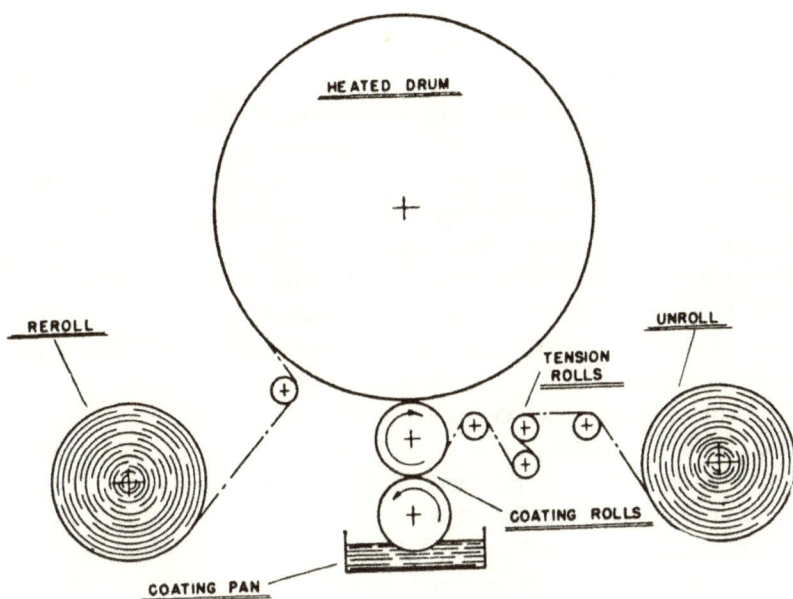

Fig. 2-4. Cast coating drum (*Courtesy of Champion Papers.*).

the outer surface of the sheet.

Three basic methods of handling the casting process are employed: (1) The sheet may be coated, and while the coating film is still wet or only partially dried the coated face is pressed against the casting surface; (2) the sheet may be coated, completely or partially dried, and then the surface rewet just before it is pressed against the casting surface; (3) the coating film may be cast on a belt or drum and the paper pressed against the wet film and dried *in situ*. The coating and paper are then stripped off together.

The different techniques are all based upon the fact that the surface of a fluid coating will conform to a surface to which it is applied and assume the appearance of that surface if the coating is dried or set and loses its flow properties. The coated paper may be caused to adhere to the casting surface (a) by the use of a rubber pressure roll, which should be preferably driven, and then held in contact with the polished face by other rolls; (b) by an endless rubber or felt blanket; or (c) in the case of a Yankee type drier, by the curvature of the drum and the tension in the paper web.

One of the major problems is to prevent any slippage or movement of the coated surface in relation to the casting surface because such slippage would tend to reduce the surface gloss and mar the coating. The problems of wrinkling or tearing of the paper at the pressure roll or on the casting surface, particularly in the lighter weight papers, requires careful control of web tension, roll alignment, uniformity of press-roll pressure, and slippage of the contact pressure belt. By driving all rolls and controlling the uniformity of pressure, however, most of the difficulties can be overcome.

Once the wet or rewet coating film has been brought up to the casting surface to receive the glossy finish, several problems may appear. Webs coated by the usual methods may carry a slight bead of wet coating on either edge, this bead may tend to build up a ridge on the casting surface and pressure rolls. Attempts to eliminate this ridge by edge doctors or by leaving a small band of paper at the edges uncoated have not generally been successful because of the tendency of the paper to stick to the casting surface where the coating feathers out at the edge. When the wet coated sheet passes under the pressure roll and makes contact with the casting surface, small amounts of air may be entrapped and carried between the coating and the casting surface. This entrapped air holds the coating away from the surface in spots, forming defects in the coated surface or in the coating itself.

Much work has been done to reduce or eliminate these effects. The trouble can be avoided by rewetting or recoating the surface of a partially dried sheet and then forcing the coated surface against the casting face while at least a portion of the coating is in a fluid state. A coating composition of high viscosity containing up to 50% of entrapped air bubbles by volume can be cast-coated in this manner without forming any defects in the paper. It is claimed that any backward flowing coating also serves to remove any particles of dirt or foreign matter that would otherwise pass through and mar the cast surface of the finished paper.

Given satisfactory coating application to the sheet, proper sheet application to the casting surface, and elimination of wrinkles and entrapped matter, the final problem is proper separation of the coated sheet from the casting surface with no marring or disturbance of the glossy surface. The adherence of a given coating to the casting surface depends on the type of metal in the final plating or the chemical composition of the surface, the presence of lubricant in the coating, the moisture content of the coating, and the uniformity, smoothness, and other characteristics of the casting surface.

In many cases where a single coated sheet has been produced by the cast process, certain microscopic and larger defects have appeared. These are varyingly described and are attributed to unequal absorption of the vehicle into the paper, unequal drying and shrinkage of the coating material, or swelling of the fibers by the water and subsequent shrinkage. Most of these problems were solved by a multiple coating procedure, where the different layers may be of the same or of different composition. In many cases the total load may be no more than is customarily applied in a single coating operation.

The usual sheet is composed of two layers, which have been generally found to produce satisfactory results. The base coating may be a paper-machine coated job, or made by a subsequent coating operation and may be air-dried, calendered, supercalendered, or cast coated. If desirable, the base coat can be colored or printed. The cast coating is laid down directly on the precoated sheet and a smooth, defect-free surface can be obtained. Several theories have been advanced as to the improvement resulting from this precoating technique, and it is believed that the absorbing power of the base coat is more even and uniform and a smooth surface film is obtained. The base coat protects the fibers from swelling when they are wetted and so eliminates defects caused by such wetting. Some authorities state that 2 lb of top coating per 1000 ft^2 will produce a perfectly satisfactory cast-coated surface.

Cast coating affords a versatility in controlled ink absorption not usually found in other high-gloss papers. In making supercalendered high-gloss papers, only enough adhesive is ordinarily used to make the sheet resistant to picking in the printing press, because with the usual calendering pressures any excess of adhesive will reduce the gloss and smoothness that can be obtained in this process. With cast coatings this is not true because the use of excess adhesive not only increases printing ink hold-up, but also enhances the gloss and improves the smoothness.

The ratio of adhesive to pigment can be adjusted to meet any requirements. Where a single coating is desirable, the adhesive ratio can be held to a minimum to maintain opacity and coverage. With multiple coating operations the adhesive ratio can be increased and some titanium dioxide pigment can be added to the formulation to produce opacity and still retain low ink absorbency. Where a colored base coating is used, the top cast coating can be made relatively transparent by using a high ratio of adhesive to pigment and by choosing pigments whose refractive indices are close to those of the adhesive (clay and casein). In such cases as much as 50% of adhesive can be used to produce the desired results.

COATING FORMULATION

Practically all the common adhesives used in other printing papers have been evaluated for use in the cast coating process. Casein, starch, soya protein, glue, and other binders have been studied, but casein is the usual adhesive. High-finishing clays, calcium carbonate, and titanium dioxides are the usual pigments. The pigments used in this process must be in an extremely finely dispersed form and extensive milling has been found generally to be a necessity.

APPLICATIONS

Glossy cast-coated paper is used as a converting base for printing and embossing for the fancy, decorative, display, and wrapping paper fields. In the graphic arts industry, where satisfactory gloss inks have been developed for its printing, it has been extremely popular.

FURTHER DEVELOPMENTS

In a new development of the cast coating process, the casting includes a protective sheath, or film, which provides scuff resistance, and adds to the luster and depth of pastels and deep colors, but the nonabsorptive surface is not as readily printable with conventional offset or letterpress inks.

Calendered High-Gloss Papers

GLEN AUGSPURGER

When it became desirable to manufacture high-gloss papers more economically and in greater widths than could be handled by flint and friction glazing techniques, supercalendering and various other smoothing processes were developed. The supercalendered papers today represent the largest volume of medium high finished product, and various modifications of supercalendered papers are extensively utilized in the high-finish field. A wide range of colors can be produced, and an increasing volume of such paper is used both in the field of graphic arts and as base papers for further specialty converting operations.

BASE PAPER

The base paper for use in this process requires a certain minimum physical strength because it must undergo at least one and possibly more coating operations and a severe polishing. A smooth, well-sized, and dense sheet, free from fuzz, is desirable since the smoother the base paper the smoother and glossier will be the surface after burnishing. If the sheet is well sized and has a relatively dense structure, the coating pigment and adhesive will remain on the surface of the sheet and less of these substances will be required to produce a given effect.

COATING EQUIPMENT

Blade, roll, air knife, spray, brush, or other types of equipment are all satisfactory for applying the base coating. The major requirement is to lay down a smooth coating that contains no surface markings and has no fibers protruding from the surface. The weight of coating laid down on the sheet varies with the end use, but heavier loads of coating generally produce glossier papers. The usual loads of coating vary from 10-20 lb (25×38—500) or to 8-15 lb (20×36—500); For a typical coat weight nomograph for a ream definition of 25×40—500, see Fig. 2-5. High or low velocity air driers, infra-red heaters, or drum dryers can be used to remove the water and dry out the sheet. The coated paper is then wound on a reel and taken to the supercalender or other burnishing equipment.

Coated Sheet Wt.,
gm./49 sq. in.

Rawstock Wt.,
gm./49 sq. in.

Coat Weight,
pounds per ream,
25 in. x 40 in., 500

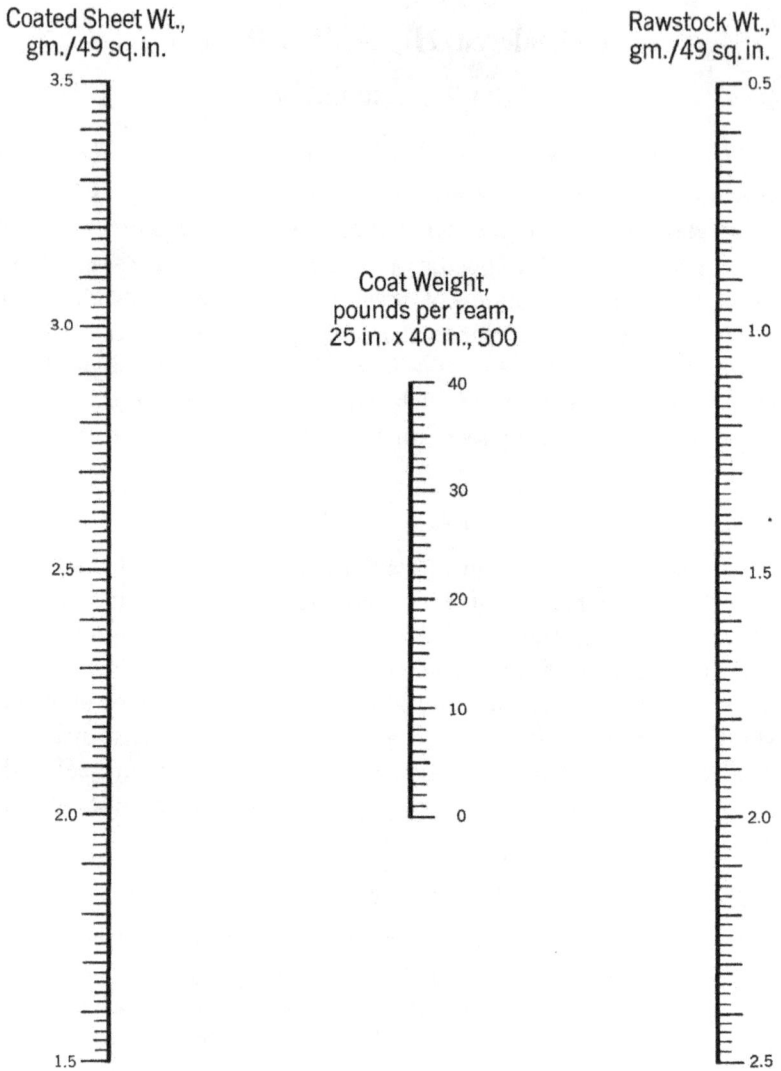

Fig. 2-5. Coat weight nomograph*

POLISHING EQUIPMENT AND TECHNOLOGY

The machine generally used to polish and burnish the coated sheets

* Davis, D. S., *Paper Ind. and Paper World,* **27,** 219 (1945).

to develop the glossy effect is the superca'ender. Such un'ts can be built to handle great widths of paper and the polishing effect on a given sheet can be varied to some extent by controlling the hardness of the cloth or paper rolls. In general cloth-filled rolls are used on coated paper. A buffing technique has also been used subsequent to calendering to increase the finish. Another technique, that is also suitable for handling great-width paper economically, has been suggested. In this method, the coated web is passed over a resilient moving bed while heated elastic blades effect the glazing action. The resilient bed may be an endless rubber-covered roll. The elastic blades are made of a material that will not mark or blacken the coated surface, —for example steel—and are designed so that they can be heated to a controlled and uniform temperature. The use of modified friction-type calenders had also been suggested, as has the use of a brush or buffing machine subsequent to the supercalendering operation.

Desired Properties in Finished Sheet

These are (1) good, uniform ink absorption and printability, (2) good brightness, (3) uniform color, and (4) high gloss. The first properties are mainly dependent upon the ratio of adhesive to pigment which should be such that the pigment is firmly bonded so that it does not pick off during the converting operation or on the printing press. The amount of adhesive should generally be kept as low as possible because in a calendering operation, everything else being equal, the smoother and glossier sheet is generally obtained with the least amount of adhesive present. The uniformity of the base paper and the coating, all else being equal, is important since if either is not uniform, the calendering operation will produce nonuniform density and surface appearance that will result in nonuniform ink absorbency.

The brightness of the final product is dependent upon the allowable pressure in the calender, the amount of moisture in the sheet, the coating formulation, the base stock, and the type of pigment used. The color match depends upon the pigments and dyestuffs, as well as the color and brightness of the body stock. The gloss is dependent upon the specific pigments used, the ratio of adhesive to pigment, the type of drive and rolls in the calender, the effect of the calendering action on the coating, the moisture content of the coating, and the pressure in the calender stack. Where the smoothing bar type equipment is employed the brightness and gloss are dependent upon the pigment used, the rat'o of adhesive to pigment, the presence of lubricants such

Fig. 2-6.   229-in. supercalender, 2000 lb air pressure (*Courtesy of Apple-ton Machine Co.*).

as talc and polishing waxes in the coating formulation, the temperature of the bar, and the pressure applied to the sheet.

Supercalender Finish

Supercalender stacks (Fig. 2-6) are probably the oldest and most commonly used apparatus for polishing paper surfaces at high speeds. In general, the finish obtained is not as high as with the other methods discussed but the cost is usually somewhat lower. Supercalendering is also the basis or preliminary process for some of the other methods, particularly brush-finishing and buff-finishing techniques.

The fundamental aim in calendering coated paper is to obtain the maximum gloss with the minimum development of color and reduction of opacity. The compacting of the mineral and fiber contained in a sheet of paper tends to reduce the opacity and the light refraction. This compacting causes a calender "blackening," which is often referred to as dinginess or calender burning. The finest calendered paper is the sheet that has arrived at the optimum point between the calender blackening and high finish.

In general the factors that result in high finish are as follows:

1. Softness of mineral pigments.
2. Fineness of the particle size of the mineral pigment.
3. Low adhesive ratio (slack-sized coating).
4. Moisture content of the paper.
5. Weight on the calender roll.
6. Hardness of the cloth- or paper-filled calender roll.
7. Temperature of the calender rolls.

In virtually all cases, all the factors that tend to increase finish also tend to increase dinginess, so that this same list could be repeated under the heading "Factors that Tend to Increase Dinginess of Calendered Paper."

In general, a moderately hard pigment such as calcium carbonate will stand more drastic calendering treatment without blackening and thus give a better finish than most soft materials such as clays, satin white, and talcs. On the other hand, the very hard pigments such as the silicates, barytes, etc. are very resistant to calender action and will result in low finishes within the practical limits of usual calender pressures.

Some pigments are too fine for practical calender processing. Carbonate precipitates have been produced that are so fine—less than 0.2 μ—that they give a very poor color and actually yield lower finishes than carbonate particles of about 0.75 μ. In general, clay particles in the range of 1μ and carbonate particles in the range of 0.75μ seem to give about the most satisfactory surfaces for supercalendering treatment. Probably 95% of the supercalendered paper produced today is coated with one or a combination of these two materials.

The smaller the amount of the adhesive used to bind the pigments to the paper, the higher the resulting calender finish. Obviously, this ratio of adhesive to pigment should not be reduced too far, for then the pigment is not sufficiently bound to stand the pick of the printing press or to resist flaking or crumbling on folding. If the paper is to be used for offset printing, the coating must be more tightly bound than if it is to be used for letterpress or gravure printing and accordingly the finish must be lower than that of the paper used in the later two processes. The nature of the adhesive has an effect on the finish, but if high-quality adhesives such as casein are used, the amount present in the paper and/or the degree of adhesion of the coating to the paper has much more effect on the finish than the normal variations between various types of adhesives. Of course, some adhesives such as certain grades of starch are recognized as being consistent producers of poor finish.

The moisture content of the paper should be kept as high as possible without excessive blackening. The actual amount of moisture varies with the nature of the pigment and adhesive. With calcium carbonate a moisture content as high as 7-8% is sometimes permissible. With some of the softest clays 4% moisture may be the maximum. The presence of satin white usually requires a lower moisture content to be able to calender without blackening.

The greater the weight per lineal inch on the calender roll, the higher the finish and the more blackening. Some carbonates and many of the barytes and silicates will stand all the pressure that is practical to exert in a calender stack without blackening. Too great a pressure per lineal inch will result in short life of the fiber rolls.

Increasing the hardness of the fiber roll has an effect that is almost identical to increasing the weight per lineal inch of the stack. The harder roll is deformed less by the steel roll and consequently the nip area is reduced. This reduction in nip area results in greater pressure per square inch in the same way as the increase in the weighting of the stack. The disadvantage of hard calender rolls is that in general they are less resilient than the soft rolls and consequently "mark" or dent more readily than the soft rolls. However, the hard rolls can stand more weight per lineal inch of nip without excessive heating than can the softer rolls.

A higher finish is obtained from hot calender rolls than from cold ones. However, practically no increase in finish is seen at temperatures above 200°F. Some calender rolls are equipped for heating with steam although generally enough frictional heat is developed in the operation of the stack so that additional heating is unnecessary. Excessive heat materially shortens the life of the fiber calender rolls.

Where a carbonate-base paper with a normal amount of adhesive (10 to 20 parts) receives a severe supercalendering with steam, a good gloss can be attained. When the adhesive is increased to 40-50 parts, however, the gloss produced by the supercalender drops off sharply. If this high-adhesive-content sheet containing a small amount of wax is calendered on a modified friction-type unit with a low friction ratio (1:1.5 or 1:2), a sheet with fairly high gloss can be obtained, particularly if the moisture content in the coating is sufficiently high during the calendering operation. With such a formulation, it has also been found possible to polish the coated surface without crushing the entire sheet.

Conditions required to produce a satisfactory finish by use of the supercalender on a carbonate and casein base sheet are given below:

Moisture content of sheet	7%
Number of Rolls on calender	9
Linear nip pressure on paper	
Top of stack	2000 lb
Bottom of stack	2600 lb
Roll density (Shore-D)	93
Roll type	Paper-filled

Buffed Finish

A need has long existed for a sheet that would have the high finish and scuff and abrasion resistance of a flint paper without the disadvantages of slow production and high cost and the general difficulties usually associated with flint papers. In a recent development, the metal-polishing operation known as buffing has been adapted to paper manufacturing. The previously coated and supercalendered web passes over a number of backing rolls. Above each backing roll is a rapidly rotating cloth-polishing or buffing roll. The buff is made up either of cloth discs or preferably of spirally convoluted cloth strips cut in such a manner that a continuous bias edge contacts the paper. An abrasive consisting primarily of stearic acid and silicate is applied almost continuously to the buff.

Other abrasives suitable for use in the process include very finely divided oxides of aluminum, chromium, iron, silicon, and similar oxides; silicon carbide; and naturally occurring siliceous matter such as diatomaceous earth or tripoli. Generally speaking, in polishing paper of white or light color, it is preferred to use an abrasive that is also light in color to avoid danger of discoloring the paper surface by particles of abrasive that might be retained in microscopic crevices of the paper surface.

The buffs are usually about 18 in. in diameter and are rotated by a

TABLE 2-1
The Effect of Surface Polishing on the Gloss of Coated Papers

Paper Sample	Gloss before treatment	Gloss after treatment
1. High-grade, varnishable, supercalendered, uncoated paper	45	70-75
2. High-grade, glossy, supercalendered, mineral-coated printing paper	63	97
3. Yellow, supercalendered, mineral-coated box liner	58	92
4. Friction-calendered, white, coated box liner	65	80
5. Cast-coated paper	83	100
6. High-grade, flint-glazed, white, coated paper	95	99

direct drive to a 1750-rpm motor giving a peripheral speed in the neighborhood of 10,000 ft min. Because of the considerable stress applied by the buff, hold-back rolls are necessary between the buffing units. Paper speeds of 500 ft/min are possible depending upon the number of buffs in the machine. The polishing roll may rotate in a direction the same as or opposite to the direction of paper travel, as desired; obviously in the former case the rotative speed must be somewhat higher than in the latter case if an equivalent differential between paper travel and roll speed is to be maintained. The paper may be repassed under the polishing roll as many times as desired or it may be passed in succession into contact with a plurality of polishing rolls.

If it is desired to polish both sides of the paper, the polishing equipment may easily be arranged so that one or more of the polishing rolls contact each side of the web as the latter makes one passage through the apparatus. The pressure of the soft resilient polishing surface against the paper surface to be polished is difficult to determine numerically. The polishing surface preferably contacts the paper so lightly as to allow considerable slippage while bringing the finely divided abrasive into engagement with the paper surface.

The effect of this treatment is to increase the gloss of a paper but is confined entirely to the surface of the sheet. The surface is closed up and compacted so that its resistance to penetration by spirit or oil varnish, lacquer or similar agents is remarkably increased, but the underlying portions and bulk of the sheet are affected little (if at all) by the polishing treatment.

Examples of the degree to which the gloss of various papers can be increased by polishing in the manner described are given in Table 2-1. All gloss figures were obtained by use of a Bausch & Lomb gloss-meter.

FORMULATION

Calcium carbonate seems to be the most popular pigment for use in these coatings because it is a bright pigment and will take a high gloss in the polishing operation. Where high-finishing clays are also used in the formulation they harden up the coating and add weight to the sheet. In some cases satin white has also been used with both clay and calcium carbonate. An enamel finish, for example, can be made by using 33 parts of satin white, 33 parts of calcium carbonate, and 33 parts of high-finishing clay. A small percentage of added satin white, clay, or titanium dioxide actually increases the gloss of the sheet over that obtained with plain calcium carbonate.

Certain high-finishing carbonate pigments, although they develop a high finish, do not retain satisfactory opacity and covering power, although this is generally a result of the average particle size or particle shape. The addition of titanium dioxide will help to overcome this deficiency without loss of brightness or gloss. Flint-glazed papers are often made with 75-100% carbonate, whereas in friction glazed formulas the usual ratio with clay is 50 to 50. The carbonate will not replace any of the talc or waxes that are used to produce slip and act as lubricants in the polishing operation. When satin white is used in the coating, it assists in waterproofing the sheet without additional treatment and it polishes to a high gloss on supercalendering or friction calendering. This pigment needs much more adhesive than is normally required for satisfactory bonding, as much as 35 to 50 parts with casein and 70 to 75 parts with starch, and this requirement is an additional deterrent to its use.

A wide range of coating weights can be used; 10-25 lb (25 × 38— 500) have been successfully handled. Casein is the usual adhesive although starch, soya protein, and polyvinyl alcohol have been claimed to be satisfactory. Starches do not produce as much gloss as does soya protein whereas casein is the best adhesive for this purpose. At least 10 to 20 parts of binder per 100 parts of pigment are generally required. The coating may be colored or tinted with dyestuffs or pigments. In general, where any substantial amount of color is required in the sheet, it is obtained by using special high-finishing pigments or dyestuffs because most pigment colors do not produce the required gloss when the sheet is polished.

Where heavy-weight coatings are desired, multiple coating may be necessary. Sheets of the highest gloss can often be produced by such multiple-coating methods, because the surface coating in this case is laid down the most uniformly and on the smoothest base.

A coated paper that has water and wet-abrasion resistance and a minimum of ink penetration is sometimes desirable. Such a coating can be produced by treating or fixing the adhesive used in the coating formulation. In some cases, urea-formaldehyde or melamine-formaldehyde resins are employed for this purpose. The waterproofing treatment may be carried out by adding the fixing agent to the coating mixture, by passing the wet or partially dried sheet through a vapor chamber, or by passing the coated sheet through a dip tank containing the fixing solution or spraying the surface as it passes through the drier. The treatment waterproofs the adhesive, which is important in papers for the converting industry, and hardens the coating and improves ink holdup, important in the printing paper field.

APPLICATIONS OF HIGH-GLOSS PAPERS

The glossy papers produced by these processes are in great demand as printing-base papers for both the converting and the graphic arts industries. The white, ivory, buff, and other delicate tints are used as printing-base papers and the colored sheets are converted and applied as box and fancy papers, greeting cards, and decorative and display papers, as well as for many other such uses. They are sold both in rolls and sheet form. The sheet makes an excellent gloss ink base and is also used in the varnishing field.

A very pronounced trend toward more attractive packaging and labeling of retail products has existed for many years. A primary cause of this trend is the "supermarket" type of retailing. The consumer is faced with a number of competitive brands of the same product, all placed on the same shelf; therefore the attractiveness of the package plays a major part in his selection. To make a package attractive, certain fundamental properties are required in packaging or labeling paper. One of the most important of these properties in certain applications is a glossy, high-finish surface. This surface can be obtained by means of a number of methods. Some of these processes not only improve the appearance of the package, but also add functional properties such as scuff resistance.

Because of the tremendous demand for high-finish papers, a vigorous effort has been made by the paper-manufacturing industry to improve the finish, reduce the cost, and increase the volume of this type of paper. The conventional methods have been slightly advanced, but what is more important, several new and improved methods have been developed or are in the process of development.

A box-cover or label paper is a functional product. It is not enough that it looks well and prints well; it must also produce a practical and satisfactory package. Some general requirements for an ideal box cover and label paper follow.

1. Abrasion-Resistant Surface. The paper must stand rough handling without scuffing or "color marking."

2. Flexibility. It must have a flake-free, leathery fold that will bend around the contours of the package and double over on itself without cracking.

3. Glossy, Nonabsorptive Surface. It must not pick up dirt readily and must resist all types of greasy finger marking. It should permit ready cleansing by wiping with a dry cloth.

4. Printability. It must be capable of good reproduction when the standard inks and printing processes are applied.

5. Appearance. The appearance must be attractive, the finish very glossy or attractively dull as desired, the colors rich and deep, and the whites bright and brilliant.

6. Handling. Box cover and label paper must handle satisfactorily for both hand and machine box-cover work.

7. Cost and Production. The cost must be competitive with that of other packaging products and the method of production capable of high-speed, quantity manufacture.

These requirements will be taken one by one and examined from the standpoint of the various high-finish papers.

1. Abrasion Resistant Surface. Supercalendered paper is only moderately resistant to abrasion. Brush-finished paper is slightly better in this respect and friction-calendered paper is considerably better. Cast-coated paper with its soft uncalendered surface scuffs and "color marks" more easily. By far the best papers with respect to abrasion resistance are flint glazed and the buffed paper. Their hard slippery surface will stand a great deal of abuse.

2. Flexibility. Flexibility is generally a function of bulk and apparent density. Usually the papers that are compacted through a calendering or glazing operation are better than those that are not so compacted and for this reason cast-coated sheets should be carefully scored. The supercalendered and brushed sheets fold somewhat better. The friction-calendered, flint-glazed, and buffed papers are the best.

3. Glossy, Nonabsorptive Surface. The nonabsorptive surface is necessary to prevent soiling of packages. Unfortunately, a nonabsorptive surface is not a readily printable surface and so absorptivity and printability are more or less opposed to each other. Those cast-coated papers that have a porous ink-receptive surface are also receptive to soiling and finger marking. Supercalendered paper is a little better. The surface is partially closed by the calendering operation. Brushed papers are noticeably better. The brushing operation greatly reduces the absorbency of the surface. Friction-glazed, flint-glazed, and buffed papers are quite resistant to soiling. They can be readily cleansed of finger marks and most normal soil by wiping with a dry cloth.

Whereas the initial appearance of the package is important, its appearance after normal storage and handling life on a shelf is even more so. The customer chooses a package by its appearance at the time he sees it on the shelf, not by its original appearance when first manufactured.

4. Printability. To produce an attractive package, a packaging or label paper must be capable of good reproduction by standard printing processes. The cast-coated papers probably stand well above the other types in printability. Their flat absorptive surface is ideal for ink reception. Supercalendered papers have good printability and will generally print gloss inks satisfactorily. Brush papers are a little less receptive to ink than are the supercalendered ones. Friction-glazed, buffed, and flint-glazed papers are relatively poor printing papers. The hard, closed, waxy surface will not absorb ink readily. The lubricating wax, which is necessary in the glazing or burnishing processes, penetrates the coating so that it is not very ink-receptive.

5. Appearance. The highest finish is obtained by buffing, flint glazing, and cast coating methods. Rich, dark colors are not common in the more ink receptive cast coated sheets. Pastels and medium shades are obtained with considerably less mottle than by the other methods, and rich, deep colors are now available in some cast-coated papers. Calendered and buffed paper does not have as good depth as the flints and frictioned sheets.

6. Handling. Friction and flint papers have a general reputation for difficult handling. Most of the other types are easy to handle. In the box cover and labeling industry it is conceded that the cast-coated papers have better pasting properties because of their less smooth back.

7. *Cost and Production.* The supercalendered, brushed, friction-calendered, cast-coated, and buffed papers are all large-scale production items. The flint papers are very definitely limited by the low speed of the flinting operation. Thus the high finished flints are only partially competitive with the other grades. Obviously, the supercalendered papers are the cheapest because of their high-speed production. The brushed, buffed, and friction papers are somewhat higher in cost— the former because of the additional brushing or buffing operation and the latter through the lower productivity of the friction stacks. The cast-coated papers are simpler to make, since they require fewer manufacturing operations. However, the initial machine installation is expensive and the production rates are low because of necessity of drying in contact with a single dryer.

chapter 3

Gift Wraps

R. H. MOSHER

Some of the earliest products of the specialty paper converting industry were the flint glazed papers used as box coverings in Germany and Austria from the late 1790's. From this beginning the industry has grown to its present proportions where the sales appeal of gift and seasonal items as well as a wide range of competitive merchandise is achieved by clever package design and the use of decorative packaging papers. Main packaging materials are available and they range in color from white through literally thousands of different shades and from a matte, dull satiny appearance to a high-gloss finish. The surface may be plain or printed and decorated with as many as twelve different colors or embossed in a wide range of patterns. If the proper effect cannot be obtained by using a paper or coated paper surface, plain or laminated films, foils, or cloth products are available to meet the exacting requirements of the packaging industry.

From the standpoint of the specialty converting industry these products can be divided into two classes: (1) the fancy and decorative box and packaging papers, and (2) the gift wraps. The first class contains products used in the manufacture of setup boxes and for wrapping packages so as to catch the public eye and serve as an identification of the product. The second group includes materials used as seasonal outer wrappings and to dress up gift items. They represent one of the few products of the industry used by the ultimate consumer as well as by the packaging industry. Such papers also find widespread use as box and envelope linings and as specialties in the papeterie field.

FANCY BOX AND PACKAGING PAPERS

The main reason for attractively packaging a product for today's market is to increase sales. If a packaged product stands out sharply on a store shelf containing many competitive products it has a good chance of arresting the attention of the passing shopper, thereby arousing interest. According to one estimate 54% of all purchases are made on the spur of the moment and more than 90% of all purchases are made by women. A suitable wrapper can be used to transform a plain product or simple box into a beautiful package that arouses desire to purchase the item. Impulse buying has become an important factor in today's marketing picture and without "eye appeal" a competitive product may find its sales lagging in an alarming manner.

In creating a paper for packaging purposes, several specific design requirements must be kept in mind. The finished paper must allow the manufacturer of the product to label or identify the contents properly, both from the standpoint of necessary sales information and to meet any legal requirements. The surface must therefore be printable and the surface decoration be so laid out that space is available to receive the necessary legend. If the product is to be a part of a merchandizing plan that necessitates affixing a label or price tag to the surface, the characteristics of the surface must be such that it will hold a gummed, pressure-sensitive, or special adhesive coated label. The sheet must be flexible enough so that it can be folded around the product and take the 90° or even 180° folds necessary in a box fabricating operation without cracking or tearing.

If the package is likely to be exposed to sunlight, it must possess fade or discoloration resistance, and the degree required may greatly influence the cost of the paper. The package surface should be more or less abrasion resistant, depending upon the expected usage, and this may be a very important characteristic if the package has reuse possibilities. If there is a chance that the package will be carried in the open where it may become wet, wet abrasion resistance and resistance to color bleeding are necessary properties to prevent smearing of the package surface and possible discoloration of surfaces with which it comes into contact. The sheet must not embrittle upon aging and must not block, either in the rolls or the sheets supplied by the coverter or in the form of the final fabricated product. The sheet must be nontoxic and should not possess any odor in its basic form, although these decorative products are often formulated so as to emit a synthetic odor such as a perfumed, leathery, or fruity smell, or other distinct odor to assist in creating an over-all effect.

The surface of the sheet should not mark or discolor upon handling, and a certain amount of water, grease, and chemical resistance may be desirable. All these properties are not necessary in every single decorative packaging paper but some are always required and others depend upon the economics of the package. As a general rule, the final choice of the sheet properties will be a compromise, and the more of these properties that can be built into a given sheet, the wider is the expected use. For many applications it is the decoration and appearance that sell the sheet, but its performance is what sells and resells the packaged product. In the following chapters many types of finishes and surface effects are described and technically classified; the purpose at this point is to outline briefly what is available in this field.

In the development of a given specialty decorative packaging or fancy paper, the converter first carefully selects the proper base paper because this sheet is the basis for all the later processing operations. The finish, color, and use of the finished product must be taken into consideration. For the usual type of setup box covering, a groundwood base paper is a common choice because the sheet is coated and thus covers up any deficiency in the appearance of the base stock. In such a case, strength and finish are not of great importance. In no case would such a sheet be planned for use in folding box, bag, or flexible packaging applications. In these cases, a kraft, sulfate-, or sulfite-base paper would be indicated because such papers possess higher physical strength and have better folding qualities. The weight of the base paper is extremely important; the usual sheet has a basis weight of about 45 lb on a 24×36—500 ream size.

The cost of the finished product will many times, and more often than necessary, govern the composition of a fancy paper. When the question of cost reduction is introduced, the manufacturer may substitute cheaper base stocks, less costly raw materials, and less expensive processing operations since the diversity of processes and the wide variety of raw materials usually permit such action. This process might involve substitution of groundwood base paper for a sulfite sheet, a friction-glazing process for a flint-glazing operation, or a lithocoated printing base sheet for a matte coated fancy paper. The amount of light-fast pigments or dyestuffs can be reduced or they can be replaced with cheaper and less resistant materials and the amount of high-priced metallic inks can also be reduced.

Water-base coatings can be used in place of organic solvent soluble varieties. Such substitutions often reduce the price by one third or one half of the original estimate, but at the same time the customer necessarily receives a reduction in finish, quality, and general appearance. Reduction is sometimes necessary, but it can turn out to be a false economy because the small savings on the unit cost of the final package may make the difference between wide customer acceptance of the packaged product and a poor position in a competitive market. The basic premise remains that the convertor must design a product to meet the specific needs and cost budget of the consumer.

Many types of papers are used in this field and a wide variety of finishes is available. The glossy papers may be flint-glazed, friction-glazed, super-calendered, or cast-coated types. The medium- and matte-finished papers are extensively used and these may also be later top-coated or varnished with casein, shellac, varnishes, lacquers, water dispersions, hot melts, or organosol or plastisol types of coatings where the specific properties that these materials offer are needed. Such coatings can be applied as clears on a wide range of colors, and where multiple coating operations are used, the pigmentation or coloring can be done in any of the applied coatings.

In the field of printed fancy papers the most popular printing technique is the gravure process, which is fairly low in cost and enables the manufacturer to bring out details and clarity in the design by using properly etched rolls. This process can be used to make many extra color shades by over-printing or using half-tone engraved rolls. Aniline printing is also popular, and although the clarity of print is not as satisfactory as with the gravure method, the cheaper roll cost and ease of multiple roll printing offer advantages. Surface printing is used to some extent and the letterpress process is useful where special effects are desired. Some lithographic or offset printed papers are produced for this trade, but the general practice is to subcontract this work.

A wide range of printed designs is used and includes large floral patterns, small flower arrangement, wide to narrow stripes, a wide variety of spotted, speckled, and irregular designs, small figures of babies, animals, Christmas figures, stars, houses, trees, birds, fish, and countless others (see Fig. 3-1). The actual printed product of any converting plant is only limited by the type of presses available and the number of color stations on the presses. The desired rolls for printing a given design can be easily procured if they can be run on the available press.

Fig. 3-1. Fancy wrapped candy package (*Courtesy of Sylvania Division, American Viscose Corp.*).

The third major method of decoration is by embossing the sheet, and a wide variety of embossing patterns is available. The designs may be exclusive or common depending upon the requirements of the package. The common designs, such as certain leather grains, moire, basket weave, swirls, and skytogen, are supplied by practically any converter since these patterns are popular and most converters own or can obtain rolls to produce the designs. The exclusive patterns, however, are the property of the manufacturer who owns or originates the design. Most manufacturers have certain designs of their own, many of them protected by the U.S. Patent Office. Where it is desirable to carry the decoration even farther, print embossing, topping, or spanishing techniques can be utilized as their application permits attractive color combinations as well as the light and shadow effect produced with embossing units.

Where the use of plain or colored coated papers, which are later printed or embossed, does not produce a properly decorative or unique sheet for some applications, other special products of the paper converting industry can be used. Films and foils, both plain and laminated to paper can be decorated in various ways to make striking and brilliant box coverings. Decorated cloth and cloth-paper laminations also can be utilized to produce attractive wrappers. Metallic coated papers in various shades of gold, silver, copper, and colors; velour or flock coatings; crystal and iridescent finishes; marble finish; mica-coated papers; and plastic coatings of many types can be used to produce a sheet that is either little or greatly different from its competition.

The question of consumer preference in design and color is one that plagues every manufacturer of fancy papers. Every company has had the unfortunate experience more than once of developing a new line of papers encompassing a number of printed patterns and a base color line of several shades, only to find that the trade almost unanimously chooses three different patterns and wants them printed upon two of the available base color shades. This problem is less acute in the development of a special paper for a given application but here again there is often a problem since in many cases the converter does not know exactly what the boxmaker will do with his product and knows even less as to what the final package will be.

Some of the larger consumers of fancy box papers are the stationery, candy, jewelry, and cosmetic manufacturers and jobbers, but large quantities are also used in general gift boxing and the packaging of any type of competitive items where women are the probable buyers. The artists and designers of the industry spend many hours and produce dozens of individual sketches before they determine the exact pattern to be the basis of the coming line. In normal times the average fancy paper manufacturer introduces 8 to 12 new printed patterns and 4 to 5 new embossing patterns per year. The older designs are retained and may be renewed again at 5-to 10-year intervals.

A recent development that has aroused a great deal of interest in the box trade is the extensive use of brand identity and trademarked papers. The use of such papers, which definitely identify the origin of the product, has been rapidly growing and has been found very successful by the large and progressive stores. In some cases the organization name is imprinted directly on the paper, and in other instances the design is used to identify the source. This type of boxpaper

Fig. 3-2.   S & S automatic tight wrapping machine for packages (*Courtesy of Stokes & Smith Co.*).

can be manufactured only in large quantities because of the cost of the art work, preparation of the printing rolls, and other special production costs.

The remaining problem confronting the fancy paper manufacturer is that of insuring that his product will handle in a satisfactory manner on package fabricating machinery.   Standard and specialized types of box-wrapping machinery are both used.   They include tight-wrapping units (see Figs. 3-2 and 3-3) where the box covering is glued to the preformed package, and specialized machinery for the production of set-up paper boxes that are covered with the proper wrapping prior to shipment; collapsible or folding boxes that are produced flat and are then set up when filled in the packaging plant; and the special types such as the bag-in-box, fitted box, and cut-out box.

Specialized package forming and wrapping units that do not involve a box as the basic unit are also finding wide application.   The package design may be simple and thus simplify the problem of the paper manu-

Fig. 3-3. S & S paper box wrapping and automatic gluing machine (*Courtesy of Stokes & Smith Co.*).

facturer, but where special folding requirements, gluing specifications, and functional characteristics are demanded, the paper manufacturer should be notified of these before the paper is produced otherwise serious difficulties may arise. These can not only cause rejection of the paper but also can slow up the entire packaging setup.

chapter 4

Decorative Papers

Many specialty papers have some functional characteristics but are basically utilized for their decorative value. Such papers are wall papers, display papers, lampshade stocks, and paper drapes. Other decorative papers should also possess some functional characteristics: those used for table cloths, napkins, and shelf papers. These sheets are utilized in the home and in business as expendable products that have a predicated life span and then are replaced by new material. These papers are generally printed in bright colors and fancy patterns and may be coated or specially treated according to their use requirements.

Domestic and Display Papers

R. H. MOSHER

SHELF PAPERS, TABLE CLOTHS, AND NAPKINS

The shelf papers, table cloths, and napkins manufactured by the specialty converting industry have been developed over a period of years. They range from the original printed tissue papers, which are still utilized as items for one-time use, to the plastic-coated, treated, and printed materials of today that can be reused in a similar manner to oilcloth or regular treated cloth. All the pigments used in either inks or coatings must be fade-resistant and must not bleed when water or alcohol is spilled on their surfaces. All ingredients should be non-toxic, odorless, and contain no components that could migrate to the surface and cause discoloration of the sheet.

The earliest shelf papers and table cloths or napkins were manufactured by printing suitable designs on plain or creped tissue, or other light-weight paper by means of gravure, aniline, or surface printing techniques. Such sheets possessed little or no reuse value and were purchased as disposable items. The inks were only partially resistant to water and most of them possessed practically no resistance to alcohol. The inks have been improved over the years so that these items now possess excellent water and alcohol resistance. Machines have been developed to print the paper, emboss the sheet in a fine pattern and then slit and sheet it to size. The printed designs cover a wide range from overall patterns to small individual designs spotted on the corners or in the center of the individual items.

The tendency has been toward more permanent papers in this field. The better shelf papers and table coverings now manufactured are based upon a strong-fibered paper—usually either a bleached kraft or a sulfite sheet that has had a wet-strength treatment or a rubber- or resin-saturated sheet with high tear resistance and folding strength. When properly treated such a sheet has the life expectancy and many of the properties of oilcloth. Another base for these sheets is the non-woven cloth now commercially produced, although this product is even more expensive than the treated papers. Napkins are manufactured from tissue and nonwoven cloth, as the heavier treated paper base products do not have sufficient softness and flexibility. The cost is also a factor in selecting the base paper.

The various papers or nonwoven-cloth bases are prepared for use by printing directly on the surface and then coating with a clear, synthetic-resin-base plastic as a protective film to facilitate cleaning, to provide grease and solvent resistance, and to prevent scuffing and abrasion of the sheet surface. Where a more light-stable colored base is desired, the sheet may be coated with a pigment composition of the synthetic-resin-base type. This is used plain, or a printed design may be applied to the surface with special resistant inks. The coatings may be applied by means of roll, knife, or gravure-coating equipment and the printing is generally done on a gravure unit. The plastic coatings may be of the solvent, dispersion, hot-melt, or organosol type. A major requirement with this type of sheet is to formulate the coating so that it will not become brittle upon aging and crack and flake off the paper surface.

In the manufacture of papers for shelf coverings, DDT, other insecticides, and fungicides are often incorporated in the coating or the base paper. These have proved to be effective and are strong selling points with such papers. In some cases, clean scents such as wintergreen, cedar, or pine, or even perfumes are added to the coating or printing ink for shelf papers designed for bathroom or closet use.

One popular shelf paper and table cloth material that is now on the market is made from a special resin-saturated base paper, printed or coated with a heavy film of a vinyl-resin organosol. This sheet possesses high strength and flexibility and excellent aging resistance. A large part of these properties derive from the relatively thick resin film on the surface. It is available in a wide range of colors and plain or printed patterns.

DISPLAY PAPER AND LAMPSHADE STOCK

These sheets are purely decorative in character and possess practically no functional characteristics. The pigments used in coatings or printing inks must be completely lightfast and bonded to the sheet by an agent that will not age and crack off. Scuff and abrasion resistance are not important for display materials, but lampshade papers do need this property to some extent.

Display Materials

The basic function of these materials is visual—the attraction of attention and the delivery of a message, either with words or by the creation of an image that puts across the basic concept by direct or abstract association. Many materials are used in this work;—crepe and other papers, cloth, foil, all types of coated and printed papers, and boards and films. The sheets may be plain or colored, coated or noncoated, embossed or printed, and include plain whites and colors, imitation leathers, printed paper with textile or wood-grain designs as well as a wide range of fancy and box paper designs.

Creped papers are widely used because of the ease with which they can be handled, their relatively low cost, and the wide range of colors and varieties in which they may be obtained. Foils and metallic coated papers are popular because of their brilliance and eye-catching color and rich appearance. Printed papers, especially lithographs, are always popular, although silk screened papers are now also extensively used.

A few special papers are produced for the display trade. Papers coated or printed with fluorescent or phosphorescent inks or coatings make displays that are startling and attract wide attention. Showcard paper must be made with nonbleeding dyestuffs or pigments in the coatings so that the water-soluble show-card inks will not bleed out from the base color. Some special papers for show-card work are made with a pressure-sensitive or gummed backing so that they can be pasted directly on the desired surface.

These materials belong to the advertising and merchandising industry and as such are constantly in a state of flux. Materials of great importance one month many be completely superseded the following. The over-all result is that aside from creped paper, which is a standby in this work, the converting industry supplies the materials required at any one time, and goes on producing these same materials for their regular uses.

Lampshade Stock

Plain or coated paper or laminated-foil or cloth-base materials are printed, embossed or otherwise decorated by the specialty converter and are then fabricated into the desired shades by the customer. The wide variety of materials used falls into two main classes: (1) translucent, (2) opaque. Transparent sheets are not generally used because they would not reduce the glare from the incandescent lamps or hide the lamp structure. The purposes of the shades are (1) decoration, (2) to cut down direct glare from the light source and either to channel it to the ceiling for reflection to the room or send it directly to the desired surface, and (3) to hide the structure of the lamp and the light source. The shade should be preferably flame-retarding in case of the overheating of a lamp.

Translucent Stock

Translucent shades are generally made from treated paper, parchment, laminated glassine, or woven or nonwoven cloth, either plain or laminated to a tissue or glassine. The base material is often impregnated with a flame-retarding compound to guard against the possibility of overheating by the lamp. In the case of laminated products, a nonthermoplastic laminant is desirable otherwise the heat generated in the illuminating system may melt the laminant. Where a treated paper is used, it is saturated with an oily, waxy, or resinous compound to transparentize or semitransparentize the sheet. Such compounds may be wax- or resin-base hot melts, solutions of synthetic resins, and in many cases, oils and petrolatum.

The base sheet is usually decorated by embossing printing with a suitable design that will show up when a light source is placed inside the shade. When embossing is the method of decoration, a roll setup to compress that portion of the sheet under the pattern without deeply embossing it is preferred. In this way, the design shows up darker when the light is turned on behind the shade, whereas the surface of the sheet still appears relatively smooth. Lightfast pigments and a binder that will not age rapidly and crack off are used for decorating shades. Surface, gravure, aniline, and lithographic printing processes have all been used, each for a specific type of design, and all have been successful.

The finished paper is generally sheeted to rather large dimensions, in terms of the usual specialty paper sizes, i.e. the sheets range from 30 in. × 40 in. to 44in. × 92 in. such dimensious are necessary because the customer wishes to have the minimum amount of waste, and small size sheets, in general, produce high waste.

Translucent shades are often pleated; for this operation the sheet must possess sufficient flexibility. The material may also be pasted or stapled to make the shade cone, perforated to take ribbons or threads, and sewed where a fancy binding or trimming is applied. The sheet must be suitable for these fabricating steps.

Opaque Stocks

Opaque shades are generally produced from coated and printed papers, foil or woven or nonwoven cloth laminated to paper, glassine laminated to paper, or film laminated to paper. In the last three cases, the paper is on the inside to produce the opacity whereas the foil, film, or cloth is the outer, visible portion of the shade. A black laminant is generally used to reduce light transmission.

1. Coated and Printed Papers. The outer surface of such sheets may have a glossy or a matte surface in a wide range of colors and may be printed with a decorative design or a pictorial representation. The finished surface is usually given a protective coating of a synthetic resin, plastic or varnish.. Instead of a plastic coating for protection, a cellophane or cellulose acetate film may be laminated to the sheet surface to provide both gloss and protection. The pigments and adhesives used in both inks and coatings must be fast to light and resistant to aging. The sheet is often treated with a flame-retarding agent to reduce danger from overheating.

Foil Laminations. Foil, either plain or colored, may be laminated to the surface of heavy paper to provide a brilliant and eye-catching shade. Dull-surfaced foil is generally used to reduce excessive surface glare. Such foil shade stock can be printed to produce the desired effect and the backing paper can be treated to produce fire-retardent properties.

Cloth Laminations. Cloth laminations are handled similarly to the foil and can be printed as desired. The finished product can be made to resemble closely expensive handmade multilayer cloth shades. The new nonwoven cloth offers some unique possibilities in these applications.

Laminated Glassine or Film. The use of laminated glassine, backed by a glossy coated paper, as a lampshade material is a recent development. The surface of the glassine is printed with an overall design to contrast with the background coated sheet. The overall sheet is opaque, even though the glassine portion is translucent. Similar effects can be produced by films laminated to coated paper, with the printing done on the reverse side of the film before lamination. The print is protected by the layer of film, and the glossy surface of the film is fully utilized.

These stocks can be embossed, but as a general rule, the natural surface is left untouched. The product is sheeted and shipped to the customer. The shade blank can be stapled, glued, pleated, or decorated and its properties must permit these operations to be easily performed.

Many complex and expensive printed papers of the brocade type and special gravure or surface printed designs are produced for lampshade applications. Metallic printing is common, and some flocked surfaces are also made. The size of the runs of individual, special or complex designs are usually relatively small, but quantities of plain colors and simple overall patterns with popular appeal can be large enough to be very desirable business.

Paper Draperies

J. C. PICKERING, JR.

The concept of making window curtains or hangings from paper is not new. However, to get a practical, decorative, and salable paper drapery, it was necessary for the paper industry to reach a point in its development where it could furnish a web that when suitably processed, could be made into such an item.

The result of development work in the paper industry was an ab-

sorbent, wet-strength web, first produced about 1930 and primarily
utilized in towels.   Such a material was capable of being plasticized,
and since cellulose plasticizers are humectants, the ability to retain a
certain percentage of strength at high humidities was vital.   Using a
well-known plasticizer as glycerin as a saturant a web could be produced that had cloth-like characteristics.   It would drape, it was soft,
had a smooth, silky "hand" or feel, had sufficient tensile strength to
be handled and could even be sewn.   Several products were made
from this paper, included bed-sheets, pillow cases, dental bibs, and
various types of medical examination sheets.   Further experimentation

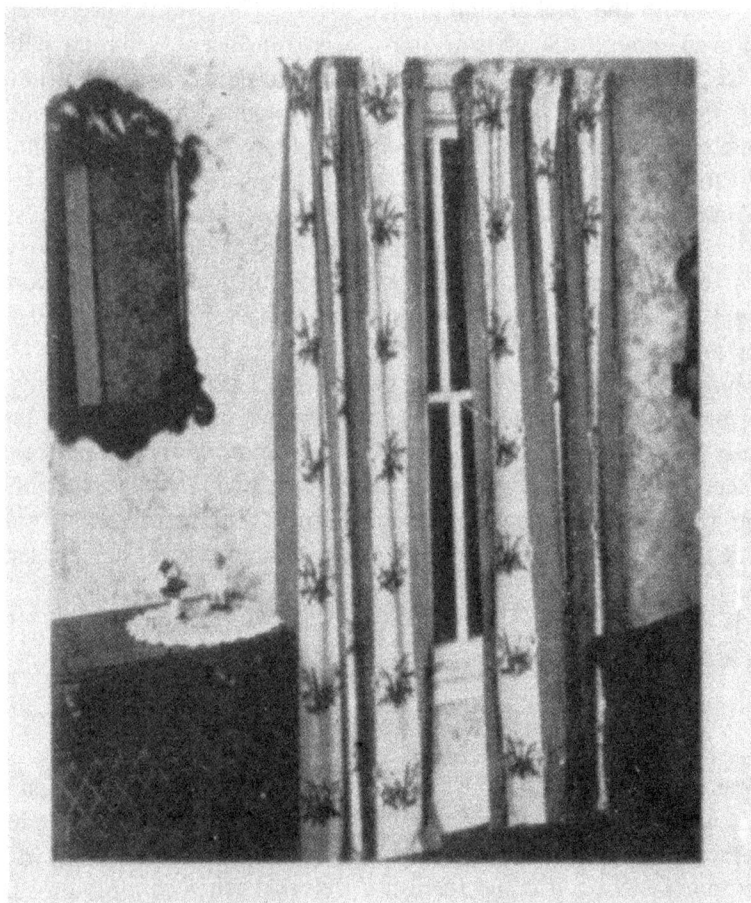

Fig. 4-1.   Paper Drapery (*Courtesy of Pervel Corp.*).

showed that a highly plasticized paper, under the proper conditions, had excellent printing properties. This immediately opened up the field of decorative papers, and suggested the possibility of a durable paper window hanging. Such a product could not be laundered or dry cleaned. It was made to be used and thrown away and replaced with other hangings at the next cleaning season—hangings of new patterns and color schemes. Such a procedure would certainly be a radical departure from the conventional scheme of reusing the same fabric draperies year after year. Another drapery product that has come into more recent use is made from the nonwoven cloths that do not require plasticizing and can be handled with normal converting equipment.

Paper draperies (Fig. 4-1) were first offered to the public late in 1938. Consumer acceptance was slow in coming, partly because of a reluctance to try anything new and partly because the word "paper" created the fear that the hanging was fragile and not durable. Yet sufficient interest was aroused to warrant the continuance of promotional sales work. With the advent of the Second World War came a shortage of fabrics and in the hunt for materials to replace cloth hangings, the public found a limited quantity of paper draperies available. As time passed and further allocations of paper for draperies were curtailed, users found that with proper care the hangings were decorative and durable. Their old paper draperies could be saved and put up for another season. The product, because of its low price and proved acceptability, soon became another one of the "nylon" items and was eagerly sought after by long queues of eager shoppers. The close of the war brought an end to restrictions on production, which had practically stopped the production of paper hangings, and manufacture was expanded as new producers entered the field and a rapid growth of the infant industry ensued.

Technology of Drape Manufacture

The manufacture of paper draperies can for convenience be divided into three steps: (1) plasticizing the base sheet, (2) printing the plasticized paper, and (3) fabricating the drape from the printed paper.

Plasticizing

This operation is usually one of impregnation, and it imparts to the base sheet the cloth-like qualities previously mentioned. The web is unwound from the roll and first passes through a bath containing the softener in water solution. The impregnated web then is passed between a pair of squeeze rolls so that the excess of plasticizer is removed. It is finally passed over drying drums where the excess of water is removed and the finished paper is then rewound in roll form. The base paper now contains the predetermined amount of plasticizer, usually calculated as a percentage by weight of the web. This process is sometimes carried out as a size tub operation on the paper machine, but it is more often run as a secondary converting procedure. Extreme care is necessary in handling a limp sheet of this kind because of the tendency to form wet wrinkles when tension is applied. The use of spreader rolls or bars normally corrects this tendency.

In the past few years, much time has been devoted to the theory of cellulose fiber plasticizing, and undoubtedly a great deal more will be given to it in the future for it is a fascinating subject and one of considerable controversy. We are only concerned here with the fact that to make a paper drape that would "drape," hygroscopic, water-soluble compounds or mixtures must be employed. The drape plasticizing agent should possess certain characteristics: (1) a narrow humectant range (2) nonvolatility (3) freedom from objectionable odor (4) nontoxicity (5) lack of color.

Glycerin has been about the best-known, most effective, and most readily available plasticizer for cellulose. It has been used in the paper industry for many years even though it has its limitations, particularly with respect to humectant range. During the second World War it was not available for use in such a nonessential product as window drapes and the search for a substitute or a combination of materials to replace it was therefore quite desperate. Since the end of the war, severe shortages and consequent high prices of glycerin have stimulated further work to find more plentiful and cheaper plasticizers. Some of the glycerin substitutes or extenders currently being used include the glycols, particularly ethylene glycol; sorbitol; invert sugar; sodium lactate; and potassium acetate.

A fairly new development has been the incorporation into the web of a flame retardent to prevent or retard the propagation of a flame. A stable and nonvolatile flame retardent will remain in the paper drape because it will not be subject to laundering or dry-cleaning.

The most logical place to introduce the flame retardent is in combination with the plasticizer during impregnation. Therefore, the substance or combination of substances used must be compatible with the plasticizer. Such a material should not volatilize or migrate from the paper, should not materially affect the stiffness of the web, and should not destroy the feel or "hand." Aluminum sulfate, ammonium sulfate, borax, ammonium and diammonium phosphate, and recently the salts of sulfamic acid (particularly the ammonium salt) have all been employed. Some of the hygroscopic inorganic salts used as plasticizers also have flame-retardent properties that make them of interest.

Printing

When the paper has been properly plasticized and conditioned, the next step is printing. One of two processes is generally used in the printing of paper-drapery material: surface printing or intaglio (engraved-roller) printing. The character of the design produced depends to some extent upon the method used. Inasmuch as the visual appeal of the product is largely the initial determining factor in selling, choice of design and coloring is of paramount importance. A paper window hanging, although it is destined for a shorter useful life than cloth, must have a design that will give the illusion of cloth and not paper.

The inks employed naturally depend upon the method of printing used. If it is surface printing, water-base inks with the proper vehicles and dispersible pigments are utilized. Many different formulas have been studied for this type of ink to obtain coatings that have a maximum resistance to the hygroscopic action of the plasticizer. Care must also be taken that the coating does not have the effect of stiffening up the previously softened base paper. In the engraved roller method for gravure printing, spirit-soluble inks are normally used. Precautions must be observed in the choice of the proper vehicle and particularly in obtaining sufficient coverage. In all cases, colors should be picked with emphasis first on light stability and second on quality and range of shades.

The design of specialized window-drapery patterns is the work of skilled artists. Available designs are limited but some variations of standard patterns are possible. Many designs are created but most of them do not get beyond the artist's sketch. In this connection, public preference is the decisive factor. If any method were developed, to determine which patterns and color combinations would sell and which would not, an enormous amount of human energy and money could be more profitably utilized elsewhere.

After a pattern has been accepted as having reasonable possibilities, the sketch is then submitted to the roll maker. The design is broken down into its component colors according to the artist's sketch, and one roller is cut for each of the colors to be used. There are four to twelve in printing a design. The completed set of rollers is delivered to the printer and the pattern must be sampled. Sampling consists of taking the artist's basic conception of coloring and then experimenting with various combinations of colors for harmony and visual appeal, until a series of colorings has been obtained, with an eye to current color trends and tastes.

Fabrication

This final part of the operation has to be dealt with as briefly as possible because of the many kinds of window hangings in rooms used for different purposes. Basically, the industry employs two methods of fabricating the printed plasticized paper into draperies: (1) sewing or stitching and (2) pasting.

Inasmuch as the paper has certain cloth-like characteristics, it is possible to sew it on power sewing machines in a manner similar to cloth. The method simulates cloth fabrication, and when properly done, gives additional strength as well as decoration to the product. Side, top and bottom hems, and rod pockets can all be fabricated by ordinary sewing. Even shirring and ruffling can be sewn into the paper to give added decorative effects.

The pasting method of fabrication requires the use of edge gumming equipment to fold over and secure with paste the side hems. Reinforcing tape and thread are often employed to give additional strength to the side hems. Rod pockets in most cases are stitched to give security to the hanging. Because the plasticizer is hygroscopic, the adhesive used must not be materially affected by moisture. Several synthetic resin glues are used.

Future Developments

In any new product such as the paper draperies, growing pains have been inevitable. Considerable work has been carried out by the industry with a view toward bettering the product. Mechanical softening as a means of rendering the sheet pliable instead of the use of a hydroscopic plasticizer is one trend of future improvement. Ways and means of incorporating more strength into the web are also worthy of concentrated study so as to produce a more durable product. This problem is, however, rather one for the base-paper manufacturer and not for the converter. Although the product has been accepted as something worth while, it will require constant diligence and continuing improvements to keep the industry a dynamic force in the paper converting field.

GIFT WRAPPINGS

The idea of decorated tissue papers for use as wrappers originated in Europe during the 17th century. From this early usage as a wrapper around women's hand boxes it has widened out in its applications until today it is widely used as a seasonal overwrapper for gifts of all types, for envelope and box linings, and as a fancy or decorative packaging medium. Well-wrapped and attractive packages have tremendous appeal from the standpoint of impulse sales, but a pretty package has also become a necessity with most gifts, and gift wrapping has developed into an art in itself. Most large stores now gift wrap packages for the customer, not only during the holiday season but at any time during the year.

The usual gift wrapping sheet is based upon a sulfite tissue paper, where the basis weight ranges from 18 to 30 lb (24 × 36—480) and the recent trend has been toward the heavier papers. Some bleached kraft and rag-content sheets are utilized where special requirements are necessary, but the ragbase sheet is rather expensive for this field of application. The heavier-weight papers, running about 24 to 26 lb are often used for gravure printing, but the lighter weights can be handled without trouble where they are desirable.

The base paper may be coated or printed and even embossed and further decorated in some special cases. The printing is done in bright or pastel colors using aniline and gravure techniques for most jobs. The most expensive papers are usually produced by the surface-printing or the silk-screen process to give the effect of handmade or block-print products. The printing may be done as a direct print or by a choke-roll process so that a solid background color is used to outline the design. The colors most commonly used depend upon the seasonal application. For Christmas the blues, reds, and greens along with gold and silver are most popular because the designs are cheery and colorful and reflect the holiday spirit. For Easter the pastel spring colors are naturally called for, the most popular shades being the violets, lavender, yellow, green, and pink. Reds and white are the common Valentine colors. Films, foils, and other special-base sheets are used.

The designs cover a wide range, depending upon the specific season for which the paper is intended. Snowflakes, Christmas trees, miniature Santa Clauses, bells, and holly are usual for the Christmas season, whereas hearts and lovers are the Valentine motif and flowers and rabbits are usually the center of attraction in Easter wraps.

These papers are generally merchandized through specialty channels that cater to the packaging trade; they are also packaged for direct sale to the consuming public by specialty houses. These companies put their offerings in attractively packaged assortments complete with wrapping paper, labels, tags, and ribbon. The packaged merchandise is then distributed through department stores, stationery stores, gift shops, and the chain stores.

Wall Papers

M. A. KUCLAR

RECENT HISTORY OF THE WALLPAPER INDUSTRY

The history of the wallpaper industry since World War II has been marked by a persistent drop in volume from a high of over 400,000,000 rolls produced in 1947 to a low of 91,000,000 rolls produced in 1963. As a result, an estimated 75% of the wallpaper manufacturing capacity has been scrapped during the two decades since World War II.

A great deal of argument has taken place concerning the reasons for this decline. Analyses in depth have been made, and reams have been written about the effect on the wallpaper market of water-thinned paints and the paint roller, the evils of pricing and poor distribution, and over-designing and over-coloring of the product. Whatever the reason or reasons behind the decline in sales, the fact remained that the consumer had become disenchanted with the product in general and had turned to other decorative media. As competition became keener, the less wary manufacturer fell by the wayside. However, those that survived emerged as stronger and wiser competitors. In recent years, a thorough reappraisal of the product has taken place from all angles: design, durability, and distribution. The results have been encouraging, and the industry emerging today bids fair to make wallcovering as popular again as it once was.

RAW MATERIALS

Rawstock

By far the largest tonnage of rawstock used in the industry today is referred to as No. 2 Hanging, in basis weights ranging from 40-60 lb per ream (24 × 36—500.) Whereas the furnish varies from papermaker to papermaker, No. 2 Hanging is usually 70-75% groundwood with supplementary amounts of semibleached kraft fibers. Probably the bulk of the tonnage consumed by wallpaper manufacturers has been treated with either melamine or urea formaldehyde resins for greater wet strength. This added wet strength was a natural outgrowth of the do-it-yourself papers that require complete immersion in water to activate the paste.

Used in lesser amounts in the industry are No. 1 Hanging and Semi-No. 1 Hanging. No. 1 Hanging is groundwood-free and Semi-No. 1 contains percentages of groundwood varying between 55-60. Some typical furnishes follow:

	No. 2 Hanging %	Semi-No. 1 Hanging %	No. 1 Hanging %
Unbleached groundwood	70-75	55-60	0
Semibleached kraft	25-30	40-45	0
Bleached kraft	—	—	60
Bleached soda pulp	—	—	20
Bleached hardwood kraft	—	—	20
Talc	—	5	10
Rosin	1.0-2.0	1.0-2.0	0.5-1.0
Alum	1.5-3.0	1.5-3.0	0.75-1.5
Wet-strength resin	0.6	0.6	0.6

Specifications

The specifications governing rawstock may be divided into two categories that overlap somewhat. First, performance specifications for the manufacturer, which aid in specifying webs that run well on the coating and printing machines; the second, performance specifications for the product in the field, which govern to a large extent the ease of installation. A typical set of specifications for a 50-lb No. 2 Hanging is as follows:

	9 oz	10 oz	12 oz	14 oz
Basis weight (24 × 36—500)	39.56	43.75	52.08	60.42
Mullen	11-13	12-13	14-15	16-17
Porosity	18-25	18-25	16-24	16-24

Tear CD-MD	35×30	40×30	60×50	70×60
Caliper, 0.001 in.	5-6	5.3-6.4	6.5-7.5	8.0-8.8
Ink Test, min	3+	3+	4+	4+
Brightness G.E.	59-61	59-61	59-61	59-61

Perhaps the most important requirement for a suitable hanging stock is that it be as uniform as possible, especially in color, texture, and sizing. In the wallcovering industry, as in no other, it is vitally important that the finished product be free from shading, because the right edge of each piece will upon installation be butted against the left edge of the adjacent sheet. Even slight variations in tone are magnified under these conditions of end use.

Pigments

As in past years, the primary white pigment used today in wallpaper is a coarse, air-floated filler clay. It fills the basic requirement of chemical inertness, low cost, and soft texture. Calcium carbonate and calcium sulfate are no longer used in any quantity because of their reactivity with the proteinaceous binder now used in the industry.

The only other white pigment employed in any quantity is titanium dioxide. Consumption of this pigment has increased tremendously since the advent of the so-called soil-proof wallcoverings. Even though the cost of this pigment is substantially higher than even a good grade of coating clay, it does possess the attribute of a high refractive index. Soil-proof coating formulations require much higher binder levels than do those for regular wallpapers. As a result, the reflective surface changes from a pigment/air interface to a pigment/binder interface. Because hiding power is a function of the difference between the refractive index of the pigment and the medium surrounding it, it follows that a pigment particle of low refractive index surrounded by a binder of nearly the same refractive index would be transparent or nearly so. The use of TiO_2 with its refractive index of 2.75 is therefore indicated.

The requirements for color pigments used are actually rather severe. Consider for a moment the case history of a wallpaper pigment during manufacture and after installation: First, it must be insoluble in water and dilute alkali to a pH as high as 10.0-10.5, for it is in this state as a coating color or printing ink. In this state, too, it must resist the ravages of sundry defoaming materials used to keep the dispersion workable during long runs. These include organic materials such as octyl alcohol, pine oil, tributyl phosphate and many others.

After the colors are applied and the sheet is dry, it is immersed in an acid/alum bath to convert the soluble protein to an insoluble aluminum salt. Here the pH is dropped to 3.5-4.0. Upon installation in an average room, portions may receive excessive heat for long periods and excessive light and exposure to the numerous greases, beverages, oils, cosmetics, and other soilants found in an average home. Truly, it is a difficult assignment for any pigment. Experience has indicated the following pigments generally serve the intended purpose:

Greens: Phthalocyanine green, pigment green B
Blues: Phthalocyanine blue, thio-blue
Reds: Red oxides, naphthol reds, cadmium red, phthalocyanine reds
Yellows: Yellow iron oxide, cadmium yellow, Hansa yellow, chrome yellows
Browns: Iron oxide brown
Black: Carbon black, iron oxide black, bone black

Pigments of Low Refractive index

These pigments as a group have two functions: to serve as low refractive index extenders in some formulations and to act as flatting pigments in others. They include: barytes, blanc fixe, diatomaceous earth, talc, and calcined clays.

Several other types of pigments in current use also deserve mention here. These include the metallics, both bronze and aluminum, and the reflective pigments—mica and pearlessence. Water-ground micas are used to provide the silky sheen effects in conventional papers, while the pearlessence pigments provide the same effects in soil-proof formulations.

Pigment Binders

The pigment binders used by the wallcovering industry today include a long list of adhesive materials. ranging from water soluble colloids such as casein and soya protein to the myriad number of latices and emulsions offered the paper industry in general. In addition, we must include various oleoresinous varnishes and solvent inks used by the rotogravure printers, who are becoming an increasingly important facet of the industry. Although there is no doubt that the synthetic resin emulsions have assumed progressively greater importance in the production of quality wallpapers, it is equally true that the largest tonnage of adhesive used today is still protein, and the bulk of that isolated soya protein.

Much work has been done and much has been written on the chemistry of soya protein as it relates to the paper industry. For information on this subject in depth, refer to the end of this section.

This chapter, however, touches briefly on aspects of the use of this material as it relates specifically to the wallcovering industry.

Soya protein, to become an effective adhesive material, must be cut (i.e. dissolved) in solutions of various alkalis. Alkalis used run the gamut from borax to caustic soda and each has something to commend it. Caustic soda and to a lesser extent soda ash give high (9.5-11.0 pH) cuts that develop maximum adhesive strength and low viscosity but are prone to hydrolysis and give poorer color. Borax is used extensively with colored pigments that might react chemically at higher pH values. Flow is invariably reduced when borax is used as the sole cutting agent. Ammonium hydroxide is employed a great deal, not only because it offers an alkalinity intermediate between borax and caustic soda, but also because, upon drying, adhesives that include ammonia become relatively water insoluble. The most water resistant coatings are made with stronger alkalis at higher pH values and are subsequently insolublized by means of solutions of aluminum acetate of varying degrees of basicity. The reaction here involves a substitution of the Al^{+++} for the Na^+ used to solubilize the protein molecule. Formaldehyde or formaldehyde derivatives are often used in conjunction with aluminum acetate to aid in insolubilization. The formaldehyde reaction is similar to a tanning reaction and is believed to result in less brittle films.

Synthetic Resins

The use of synthetic resin emulsions began with the introduction of styrene-butadiene emulsions shortly after World War II and has grown phenomenally since—both in volume and scope. Appearance of these soft resin emulsions allowed wallpaper chemists to formulate coatings at higher binder levels without the brittleness inherent in coatings high in protein. In effect, they opened the door leading to development of soilproof wallpaper. Although resin emulsions are no doubt being used as the sole pigment binders in some formulations, it is probable that the majority of formulas contain mixtures of protein and a compatible emulsion. A brief run-down of the various types of resin emulsions in current use follows:

Styrene Butadiene

This material unquestionably offers the greatest value per adhesive dollar spent, if one can overlook the fact that it tends to eventual embrittlement and a slight weakness in light resistance as the butadiene oxidizes upon exposure to air. However, it blends well with soya protein and provides good wet rub resistance and good flexibility in coatings where it appears. In printing colors, it tends to foam badly and is seldom used.

Polyvinyl Chlorides

These resins are usually copolymers of polyvinyl chloride and polyvinyl acetate or vinylidene chloride. They are usually supplied as plasticized systems, with plasticizers ranging from esters and phosphate types through internally plasticized materials containing acrylonitrile. As a group they offer greater flexibility in formulation than styrene-butadiene and exhibit better resistance to aging and sunlight. Their strong point, however, is increased resistance to solvation by greases and oils, a prime requisite for soil-proof wallpaper adhesives and coatings. In cost, these resins are somewhat higher than styrene-butadiene.

Polyvinyl Acetates

These emulsions may be had as straight polyvinyl acetate resin emulsions of varying molecular weights and as copolymers with a variety of materials including acrylates, chlorides and maleates. They offer improved resistance to yellowing upon exposure to light and heat and great flexibility in formulating coatings, although they tend to hydrolyze somewhat with high pH dispersions of soya protein. Coatings prepared from polyvinyl acetate emulsions generally have less wet rub resistance than those made with polyvinyl chloride and styrene-butadiene resins unless special precautions are taken with the protective colloids and surfactants found in these systems.

Acrylic Emulsion

From the standpoint of quality, the acrylic emulsion probably has the most to offer the wallpaper industry. Its stability to exposure to light and heat is unsurpassed. It does not have the hydrolytic tendencies of the polyvinyl acetate, and formulations prepared with acrylic emulsions may be made as durable as any yet offered. Although acrylics have made inroads into the paper trade, their higher cost has prevented them from gaining wider acceptance.

Solvent and Oleoresinous Binders

These materials need only to be touched on here because their use is still rather small and where employed they are bought as finished inks and are not prepared by the wallpaper manufacturer. Oleoresinous inks at one time were used extensively by wallpaper manufacturers when engraved roll printing was widely practiced. Papers made by this method fell into disfavor prior to World War II and most of this equipment was consigned to the scrap heap. Some volume of wallpaper still is made this way, but has largely been restricted to wood grain effects. Inks in this category are characterized as having low pigment solids and are relatively transparent. Drying takes place partly by solvent evaporation and partly by oxidation.

A "solvent" ink in the wallcovering industry is almost always a pigmented solution of a vinyl resin copolymer in a solvent such as methyl-ethyl ketone, methyl isobutyl ketone, or isophorone. Drying is by evaporation of solvent only. Increasing quantities are being used in the wallcovering industry since the advent of wallcoverings based on vinyl coated fabric and vinyl film/fabric or film/paper laminates.

METHODS OF MANUFACTURING

Although the manufacture of wallpapers since 1955 has remained basically the same insofar as the coating and printing are concerned, several operations have been added. They include prepasting, trimming, packaging and—increasingly—an operation loosely referred to as plastic coating (Fig. 4-2).

The first step in the conversion of design into a wallcovering is to transfer an artist's ideas to a print roller. Almost all wallpaper and a large share of vinyl and fabric wallcoverings are today printed by the raised surface printing method. The remainder of the wallcoverings produced are made either by engraved roll printing and, to a lesser degree, by the silk screen method. Since gravure and silk screen printing are used so little for wallpaper manufacture, this discussion is confined to surface printing, the method that is peculiar to the wallpaper industry.

Much progress has been made in the field of print-roll making since the era of the wooden print roller. These rollers, with the pattern tediously copied by hand onto a wooden cylinder and subsequently outlined with thin brass strips and filled with felt, left much to be desired as a manufacturing implement. The wooden cylinders had a

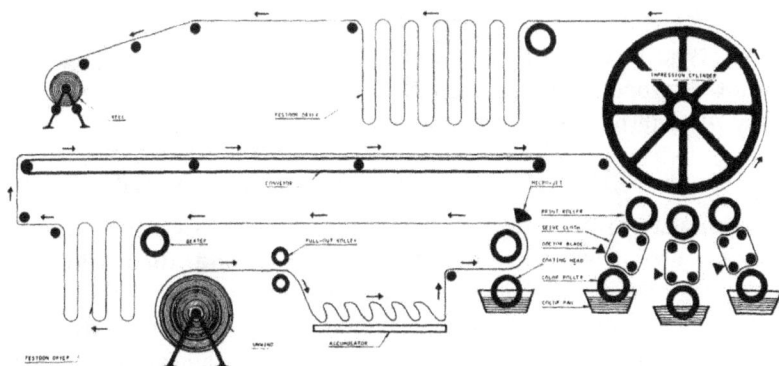

Fig. 4-2. Schematic diagram of wallpaper production line (*Courtesy of the Birge Co., Inc.*).

tendency alternately to swell and shrink from the moisture in the printing colors. The felt printing surfaces also tended to dimensional instability and soon became hard because of accumulations of residual colloid and pigment. At their worst, the cylinders would warp badly and sometimes split apart. Such was the picture when cast aluminum alloy rollers were introduced to the industry. Although the design still had to be transferred to the roller by hand, the cutting of the pattern into the metal by use of routing tools was much faster than the brass-and-felt-inlay method heretofore used. Certainly the finished roller was more stable to moisture than the wooden roller. Some changes had to be made in printing color formulation to correspond with the new printing surface, but once this was accomplished the industry turned almost entirely to routed metal rollers.

In recent years, a method of acid etching designs into magnesium sleeves has been perfected. These sleeves, subsequently fitted with ends, and chrome plated, have proved to be a tremendous improvement over the routed rolls. They have the advantage of being able to reproduce intricately detailed designs at far less cost than the routing process. Although simple designs can still be produced more cheaply by the routing technique, the acid-etched roll will ultimately dominate this field. All that remains is to devise a method to transcribe to a print roller separate colors directly from an artist's design. A process of this type is thought to be near perfection in Europe at present. When this process is made available, the tedious job of print-roll making finally will have been made a science rather than an art.

COATING, PRINTING, AND CONVERTING

When a set of print rollers comprising a design is completed and is ready for use, it will be sent to the print room and fitted with shafts and a gear of a size to correspond with the repeat of the design. The impression cylinder may have as few as four or as many as twelve printing stations spaced under the bottom half. The top half is unencumbered so it can be raised to separate the impression cylinder from the print rolls for threading of the web or for idling periods. Register is controlled by gears on the shafts of the print rollers fitted to a master drive gear of the same size as that of the impression cylinder. The impression cylinder is usually covered by a rubber jacket, although cloth-lapped cylinders are used by some manufacturers. Color is furnished from a color pan to the print roll by a felt sleeve and web tension is controlled either by tension rolls or by suction boxes.

While the print rolls are being installed and fitted in the printing machine, the colors for the design are being mixed. From a palette of 12 to 16 colors, the colormixer not only matches by eye the various shades in the design but mixes very close to the proper amount of printing color necessary for the given run. When it is realized that a pattern may contain 12 different shades, and the usage of these colors may vary from 1-15 gal/hr, one may appreciate the skill and experience necessary to perform this task.

An additional factor not often considered is that a colormixer must never waste any tailings left over from a previous run. These are always saved until they can be converted and used on another pattern. It is this factor of matching colors with tailings from previous runs that has precluded any attempts made toward applying instrumentation to this facet of wallpaper manufacture.

After all colors have been matched to the supervisor's satisfaction, and the print pattern is in register, actual production of wallpaper begins. In this country, nearly all wallpaper is made in what is referred to in the trade as "tandem" operation. The web of rawstock is fed into the process and is machine-coated with an opaque ground, dried, printed, and dried again and usually plastic-coated without the web ever being broken. Some producers have even added embossing and cutting operations to this sequence, all without once breaking the web.

Tandem operations have certain advantages, especially if the manufacturer's method of operation involves long production runs of a given pattern. Low wastage and a small labor force are required with this system. Conversely, certain advantages accrue when short production runs are scheduled. Because a relatively large amount of time is necessary to install a pattern in a printing machine, all the supporting equipment must be idle while this is being done.

After the web has been coated and printed, the entire face is over-coated with a conversion bath consisting of aluminum acetate, either alone or in combination with a nonionic resin emulsion. The aluminum acetate serves to insolubilize the protein fraction of the binder, whereas the resin fraction lays down a clear, hard film over the surface. When this coating has dried, the usual practice is to put up the web in reels and to transfer them to a finishing department for embossing and trimming and cutting into double or triple rolls.

EMBOSSING

Most of the embossing in the wallpaper industry today is simple pressure embossing, which employs a steel roll and a paper bowl made of paper discs compressed on a shaft. The rolls are pressed tightly together and the wallpaper web is run in between. Although this is a suitable technique for very fine grained embossings, the manufacture of more heavily embossed papers requires the paper bowl to be geared to the steel roll and run in while wet, thus tranfering the mirror image of the pattern to the paper bowl. In effect, this procedure creates a matched set of rolls. Quite often a web exposed to an exceptionally deep draw emboss must be steamed as it passes through the nip to soften the sheet and reduce the tendency to split.

TRIMMING

Concerning trimming of wallpaper, not much can be authoritatively stated. Probably as many methods of trimming are used as there are manufacturers. One thing all have in common, of course, is that the trimming must be done at the same time the pattern is printed to insure accuracy. In some mills the edge is fully trimmed right at the printer; in others, it is only score-cut. In still others, a trim line is printed on the selvage in register with the pattern. This line is subsequently picked up by an electric eye that guides the trimming knives or may guide the web into stationary knives. Obviously, the longer the trimming operation can be delayed, the better the process, because the selvage serves to protect the edges of the roll during finishing and final processing.

PREPASTING

The use of prepasted wallpapers has experienced a slow but steady growth in the years since World War II. The very first of these papers had starch/dextrin remoistenable gums as pastes and, although they were satisfactory when freshly coated, the starch had a tendency to retrograde and to lose solubility upon aging. Perhaps the greatest shortcoming of these early papers was a weakness in washability that caused the colors to run when the sheet was immersed in water to activate the paste. Not until pretrimmed, plastic-coated and prepasted papers were introduced was the prepasted wallpaper idea really launched.

Along with this package came the idea of the waterproof cardboard water box. This was of a size to hold a rolled strip of paper and eliminated the need for the amateur decorator to wet the strip in a bathtub of water and carry the strip dripping through the house to the point of application.

In vogue today in this industry are two distinct prepasting processes. Probably the largest volume of prepasting is done by means of the dusting process. This entails wetting the wire side of the web with water thickened enough to hold it at the surface for a few seconds. The web then passes through a duster that deposits a film of dry, blended ahesive on the back surface while still wet. In some mills this operation is performed after printing and in others it is performed on the rawstock. This process has the advantage of great flexibility in formulating the dusted adhesive and in quick drying. Its prime disadvantage is in the dust and dirt it generates in subsequent processing. In addition, the dusting process does not lend itself readily to control and the papers made this way often carry an overabundance of paste to compensate for normal fluctuation in the operation.

The second method is the liquid prepaste method, wherein the ingredients are dissolved and applied on conventional coating equipment. This method entails the removal of large quantities of water, because wallpaper pastes by nature are high-viscosity and low-solids formulations. While good control of coating weights is possible with this method, paper coated with this type of material tends to curl badly if overdried.

A third method of prepasting that is an attempt to circumvent the failings of the two previously mentioned processes has recently appeared. It involves a two-step process of the *in situ* formation of a sodium polyacrylate adhesive possessing the characteristics desirable in a remoistenable wallpaper paste. Step one consists of a coating to render the back of the sheet alkaline; step two is the deposition of a coating of alkali-soluble acrylic resin. This method is too new to assess completely at this time, but shows promise of overcoming most of the failings of the two older processes. It allows a great deal of latitude in formulation, is relatively inexpensive, and the coating can be dried quickly—three prerequisites of a successful prepasting formulation.

TRENDS IN THE INDUSTRY

In an effort to rebuild the lost wall-covering market, wallpaper manufacturers have come up with a welter of ideas to improve their products. Most of the earlier efforts were aimed at improving the durability of the product; later came the do-it-yourself papers with enhanced ease of application and, lately, efforts have been made to facilitate ultimate removal of the wall coverings. Several constructions have gained acceptance in the field and are noted here.

Nonwoven Webs

Among the first efforts to improve the strength characteristics of wallpapers was the introduction of wall coverings printed upon nonwoven fabric substrates. The early nonwovens used for this purpose were made mostly of cotton fibers with acrylonitrile binders. They were made into wall coverings with greater tensile, edge tear, and wet strength than those of the conventional wallpapers and immediately became competitive with existing woven fabric wall coverings. Subsequently, the paper industry regained the initiative with webs of this type made on paper machines equipped with rotoformers and in-line saturating equipment. Webs of this type have the advantage of flexibility of construction. Any combination of fibers and saturants can be used to give varying results, and the formulator is in a position to take advantage immediately of new fibers and resins as they appear on the market. In addition, wall coverings of this type may be formulated with high MVT rates to produce "breathable" wall coverings.

Vinyl Laminates

As the price of extruded vinyl film dropped after World War II, a number of manufacturers began experimenting with vinyl-to-paper laminates. The early products required special quick-tack adhesives for hanging, because of a persistent curling when the paper side was wet with paste. (Paper expands slightly when wet and vinyl film does not.) Later constructions of this type use higher sizing in the paper stock and a softer, more extensible vinyl film to relieve the problem somewhat. These products as a class suffer from a tendency to delaminate at the paper/vinyl interface and are prone to attack by mildew because of a very low moisture-vapor transmission rate.

Scrim/Paper Laminates

A third web construction offered to the trade is a laminate of a fairly low thread count scrim and a regular hanging stock. In this construction the scrim is applied to the wire side of the web and the felt side is coated and printed. The scrim facilitates removal of the wallcovering when the time comes to redecorate.

In addition to the various web constructions now being offered, progress in the wallpaper industry is being made in other directions. Soilproof wall coverings are appearing in increasing numbers, and even the cheapest papers now bear the legend "plastic coated." Improvements have been made in fade resistance of pigments used and in resistance to yellowing of resinous binders. Nontarnishing metallic pigments appeared in wallpapers in 1964. The industry will soon reach its goal of making papers that will be easy to apply, of providing serviceability during the lifetime of the product, and of achieving easy removability when the time comes to redecorate.

Miscellaneous Decorative Papers

R. H. MOSHER

Several types of decorative papers produced by the specialty converting industry do not fit into any of the other larger groups of classification. Such papers are used in one or more specific applications and are manufactured by a special technique or by using a type of material that is not common to the rest of the industry; therefore they will be discussed separately.

Such papers are: luminescent coated and printed papers, metallic coated papers, mica-coated papers, dull-coated papers, decalcomanias or transfers, flock-coated or velour papers, marbled papers.

LUMINESCENT COATED AND PRINTED PAPERS

Luminescent pigments can be fluorescent or phosphorescent. Fluorescent pigments emit visible light only while they are exposed to an exciting light source, which is usually of the ultraviolet or "black-light" type. Phosphorescent pigments also emit visible light when exposed to ultraviolet rays, but their main advantage is their ability to continue to emit visible light after the exciting source has been removed. With phosphorescent pigments, daylight or visible light can also be used as an exciting medium.

The fluorescent pigments possess daylight colors of varying shades of yellow, blue, gray, cerise, blue green, and white. When excited by "black light," they exhibit colors that range from orange red to yellow, green, blue, and blue-white. The pigments are fairly coarse and can be formulated for use in coatings or in inks.

The phosphorescent pigments can be divided into two main classes: those with a short afterglow period (30 minutes to 2 hours) and those with a longer period of afterglow (6 to 12 hours). The short afterglow types are very coarse in texture, and efforts to reduce the particle size generally result in a degradation of the light properties. These pigments can be obtained in green, orange-yellow, blue-green, and blue in both the daylight and the excited state. They can be used in the formulation of coatings and printing inks. The longer-afterglow types are also quite coarse and diminution of particle size will cause degradation in properties. They are both very susceptible to moisture and cannot be used in a water system. Their use is limited to coatings and inks in which organic solvents are the basic carriers.

The fluorescent pigments are usually based upon zinc sulfide or zinc and cadmium sulfides. The phosphorescent pigments are formulated from calcium and strontium sulfides. In general practice, small amounts of other metals such as copper, bismuth, silver, or manganese are mixed with the pure soluble compounds as none of them alone would have a brilliant luminescence. This activating metal causes the pigment to glow colorfully. Both the phosphorescent and fluorescent pigments can be modified in their daylight form by the addition of small amounts of dyestuffs or suitable pigments without materially affecting the luminescent color or brightness. Fluorescent pigments can be used in casein, starch, or water-dispersed resin systems. Varnishes or lacquers are also satisfactory vehicles for more resistant films.

The intensity and length of afterglow of phosphorescent pigments depend upon several factors: (1) type of pigment, (2) medium in which it is incorporated, (3) intensity of exciting light, (4) length of exposure time, and (5) temperature. The amount of activation required to obtain maximum afterglow depends upon the type of pigment. In general, a pigment with a long afterglow requires more exposure time than a pigment with a short afterglow.

Many dyestuffs can be used in similar applications although those commercially available at the present time are limited to fluorescent types that require excitation before they show any·visible radiation. These materials are available in a wide range of daylight colors and solubilities and a correspondingly wide range of fluorescent colors. They find a limited use in both coatings and printing inks.

The coatings and printing effects obtained with these materials open up many interesting applications. Treated paper can be used in safety papers, decalcomanias, match cover stocks, greeting cards, gift wraps, synthetic leathers, adhesive and marking tapes, and religious item mountings. The printed applications include special wallpapers, paper drapes, shelf papers, gift wrappings, lampshades, and display and advertising papers.

METALLIC COATED PAPERS

Some of the most popular fancy papers, which have been on the market for many years, are the metallic coated papers. Such sheets, produced on a wide range of basis weight papers, and with gold, copper, silver, or colored metallic surfaces, are utilized in many ways in the box and fancy paper field for match covers, menu and brochure covers, greeting cards, tags, labels, announcements, booklet covers, display papers, and many other applications.

There are three types of metallic coated papers:

1. The standard metallic coated papers made by dispersing a gold, copper, aluminum, or other colored metallic pigment in an adhesive vehicle and coating this on a sheet of paper. The pigment is firmly bonded to the sheet surface and the resulting product gleams in a

brilliant and truly metallic manner.

2.   A second type made by coating the paper with a tin, lead, cadmium, antimony, or other metal oxide dispersed in a suitable binder and friction-calendering the resulting sheet.   This operation burnishes the relatively dull coating so that the resulting sheet has an appearance in many ways similar to a metal foil paper.   The sheet may be colored to represent gold or copper by incorporating dyestuffs into the coating or by staining the finished sheet.   In many cases the product is coated with a clear lacquer to enhance the gloss and protect the surface from tarnishing.

3.   The third type is prepared by a dusting technique.   The paper is first coated with an adhesive layer, which is then dusted with the gold powder.   The excess powder is removed and the sheet calendered to produce the desired effect.

Such papers can be used as base papers for printing or can be laminated to the surface of other papers or boards for decorative applications.   Some of the aluminum-coated papers have been used as structural insulation because of their ability to reflect heat and light waves and prevent the loss or gain of heat through a wall by radiation.

Technology of Metallic Coatings and Coated Paper

Pigment Coated Metallic Papers

The general technique for producing metallic coated papers involves the dispersion of the metallic powder in a suitable vehicle and the application of this coating compound to the paper surface.   Any desirable base paper can be used, and it may be precoated if necessary.   The paper must always possess uniform formation, density, and surface finish and should be fairly highly sized to prevent penetration of the coating adhesive into the base paper.   The base coating may be plain or pigmented, and can be formulated from casein, glue, starch, or synthetic or natural resins.

Metallic coatings may contain bronze powders (gold or copper), aluminum powder (silver), or the various specially colored metallic powders.   These metallic powders vary in particle size, particle shape, degree of polishing, and density.   The specific requirements depend upon the adhesive and vehicle formulation used and the method of application.   Some powders are highly *polished* and contain a high percentage of stearic acid, olive oil, oleic acid or other polishing medium and others are low in polishing agent.

The reason for incorporating a polishing agent is to produce "leafing" of the metallic particles in the vehicle after it is laid down on the paper. To produce a high metallic lustre, the flakes must go to the surface of the film and so orient themselves that a continuous film of metal is obtained. When "leafing" metallic powders are applied to the sheet in a vehicle that permits this phenomenon to take place, the flakes are carried to the surface and form the desired film. Leafing is a surface-tension phenomenon, and the higher the surface tension the more rapid and complete is the leafing. The coverage of these coatings depends directly upon the particle size (i.e. the finer the particle size the better the coverage), smoothness, and hiding power. The metallic brilliance of the resulting coating varies inversely with the particle size, i.e. the larger the individual particles the greater the metallic brilliance.

The adhesive and vehicle used produce varying effects, and these are selected according to the type of metallic powder used. With vehicles dispersed in water, such as casein, glue, starch, soya protein, or natural or synthetic resin or rubber, relatively coarse golds are used containing some polishing agent or agents that produce water miscibility. These coatings can be applied to uncoated base papers by means of brush, roll, spray, or air-blade coaters. Synthetic-resin or rubber-base vehicles, the so-called plastic coatings, which require the use of solvent or hot melt application techniques, generally employ finer, highly polished metallics; these are laid down on precoated base papers.

Roll, gravure-roll, or knife coaters are generally used as the application equipment. The type of adhesive has much to do with the final properties of the coated sheet. Water-base metallic coatings are generally more brilliant, require heavier loads for coverage, and do not possess much film flexibility. The plastic-coated metallic papers show less brilliance and brightness, but they are more flexible, require less of a coating load to obtain coverage, and show a smoothness and luster unobtainable by other means. The water-base coatings tend to tarnish faster than the plastic types.

The finished coated paper is usually calendered or embossed to bring out the full brightness and luster of the metallic surface. In some cases the coating is brushed to obtain special effects.

Synthetic or Frictioned Metallics (Argentine)

The so-called "silver paper" is made with a metal oxide base coating (Argentine) that was originally produced from a mixture of zinc and antimony oxide powder. In later formulations tin oxide, cadmium oxide, lead oxide, and other waste or by-product metallic oxides were also used. These compounds are mixed with starch, casein, glue, or synthetic or natural resin adhesives and coated on the paper in a manner common to other pigmented coatings with water as the carrier. Brush, roll, spray, air-blade or other special application equipment are all satisfactory for the job. Almost any type of coating base stock can be used, but the sheet should have uniform density, formation, and sizing, and as smooth a surface as possible is desirable. The coating is applied in a thin layer to the paper surface; it is possible to obtain a well-covered, readily frictionable surface without the use of excessive amounts of polishing waxes. If casein is used as a binder no further addition of frictioning agents is necessary, but if starch, glue, or resins are used, some polishing wax in the formulation is generally recommended for best results. The coated paper can be dried in the conventional manner and then friction-calendered to produce the desired smooth silvery metallic finish. Synthetic golds and colored metallics can be produced by the addition of dye stuffs to the coating mixture or the finished frictioned sheet can be stained in a final operation.

A typical formulation is

	Parts
Metal oxide	100
Casein	16
Polishing wax	10

Two layers of the metallic film should be applied to a pigment-coated base stock. A pyroxylin top coating is suggested to improve the overall appearance and properties of the sheet.

Dusted Metallic Papers

Dusted metallic paper can be produced in a wide range of basis weights because the finished sheet only requires calendering or embossing to be ready for use. Almost any type of base paper can be used. The equipment required is rather unique, however, although very similar in design to that used for producing flocked or velour and abrasive papers. A roll coater is generally used to apply the adhesive to the sheet, and this is followed by a duster, excess powder removing attachment, and rewind unit. The finished sheet is then calendered or embossed to set the metallic film and produce the desired smooth and brilliant surface. This operation also tends to adhere the powder firmly to the adhesive layer. The finished sheet may be lacquer-coated to protect the metallic surface from tarnishing.

The adhesive used to make the metallic pigment adhere to the paper may be shellac or another natural resin, a synthetic resin, a rubber, or a protein such as casein, zein or corn protein, or soya protein. The formulation may be solvent-, heat-, or pressure-sensitive. The solvent- and pressure-sensitive coatings must be coated, then dusted, brushed, and the excess coating mixture and any solvent must be removed before the paper can be rolled up. The heat-sensitive coatings can be laid down on the paper, dried, and cooled. Then the sheet can be wound up in one operation and dusted in a secondary operation if desirable.

Such a system is described by one author as follows: The paper is coated with shellac dissolved in denatured alcohol by means of a roll coating unit, then passed over a steam drum and dried, and finally wound up. This sheet is then dusted with bronze powder in a second machine, following which the paper passes over a heated surface to soften the shellac so that the bronze will adhere firmly to it. Two rapidly rotating rollers covered with flannel rub the bronze firmly into position and dust the residue cleanly from the paper surface. A suction box is disposed above these rollers and this with an air exhauster unit collects the freely flying bronze for reuse. The excess bronze still on the paper is then rinsed off with water and the wet paper passes over a drying drum that dries off the water and, by softening the shellac layer, again causes a more complete adherence of the metallic powder. The sheet can then be calendered, friction calendered, or glazed, and a thin lacquer coating can be applied. Such papers probably show the most metallic properties since there is no binder on the surface to dull the metallic finish.

MICA-COATED PAPERS

Mica-coated or satin papers are a special type of fancy and decorative sheet whose surface glistens and sparkles with thousands of tiny reflections of light. The sheet is produced in a wide range of colors and may be printed or embossed at will. The most popular colors are the pastel and light shades and are used in the fancy box and packaging field for labels and various other decorative applications. The sparkling effect is obtained by suspending finely ground mica flakes in the coating mix. These orient themselves in the coating when it is laid down on the paper so that their reflecting surfaces are parallel to the surface of the sheet, with the result that they act as countless tiny mirrors that reflect the light. Mica is usually the only pigment used in these formulations as it is not desirable to cover the brilliance in any way. No colored pigments or pulp colors are generally used in the coating to obtain colors for a similar reason, but dyestuffs are commonly employed to attain the desired shade.

Base Paper and Coating Equipment

The base paper for mica coatings is similar in basis weight and structure to that used for dull coats or other box papers. Its basis weight ranges from 38 to 65 lb (25×38—500) and the sheet should be well-sized and of uniform formation and density with a smooth surface finish. The coating can be applied by brush, roll, spray, or air-blade coating equipment, all of which produce satisfactory sheets. The only major problem is to keep the mica flakes suspended in the coating mixture prior to application to the paper and to obgain a good, uniform coating on the paper where the flakes are oriented in a plane parallel to the paper surface. Light coating loads (2 to 4 lb, 20×26—500) are generally used.

The coated sheet is usually given a calender finish after coating and is often embossed before shipment.

Coating Formulation

A typical mica coating formula is

	Parts
Mica flakes	55
Casein (dry)	14
Dyestuffs	to obtain desired color
Suspending agent	50

Dispersing agents for the casein, antifoam compounds, leveling agents, plasticizers, and possibly some polishing wax such as carnauba should also be included. As a suspending agent a gum, starch, water-soluble resin or other compound is used to assist in maintaining a good dispersion of the mica prior to coating. Waterproofing agents can be incorporated if this is desirable.

In case of a demand for certain special coating effects, mica flakes may be incorporated into pigment coatings, but this is not a regular practice because the effect of mica is very much reduced by the presence of the pigment.

DULL, SUEDE, OR MATTE COATED PAPERS

Dull, suede, or matte coated papers are used widely in the fancy, box, and decorative paper fields because of their diffusely reflecting, dull, satiny surface. Although they do not possess the striking appearance of the highly glossy papers, they do exhibit a soft, rich effect that is widely utilized. They are produced in a wide range of colors and can be made lightfast and water and wet-rub resistant. The papers print well, although they usually tend to absorb more ink than do the glossy coated papers. The commonly used white pigments include carbonates, clays, and blanc fixe. Pulp colors and dry pigments are both utilized to produce the desired colored shades. The sheet is generally given a light calender treatment to smooth the coated surface but not enough to produce any gloss or finish. Casein is the adhesive commonly used, although some of the new synthetic resins have shown definite promise. Color and brightness are extremely important, particularly in the light and pastel shades, and any dark color contributed by the adhesive tends to reduce both brightness and purity of shade. The major problem with this type of coated paper is to obtain smoothness without gloss.

Base Paper and Coating Equipment

The base paper generally used is a soda-sulfite sheet, ranging in basis weight from 38 to 65 lb (25 × 38—500). The paper should be well-sized and show a uniform density and formation with the surface uniform but not too highly finished, as a slight "tooth" will assist in producing the desired dull effect.

The paper is generally coated on brush, air-blade, or spray equipment, all of which produce equally good sheets. The main requirement is uniform color distribution on the paper to produce uniformity of shade and coverage on the backing sheet.

Coating Formulation

A few standard formulations for four shades of colored dull coats are listed here. The color pigments are given on a dry basis, but both dry pigments and pulp colors are commonly used. The amount of adhesive is generally about 16 to 20%, based on the adhesive-pigment formulation. Other agents added to the formula, besides those required to solubilize the casein, are wetting agents, dispersing agents, leveling agents, antifoaming compounds, and such compounds as

Dull Coat Formula

White	Parts	Yellow	Parts	Blue	Parts	Maroon	Parts
Casein	225	Casein	100	Casein	264	Casein	81
Special coat-		Special coat-		Special coat-		Maroon	257
ing clay	365	ing clay	283	ing clay	113	Red	6
English clay	554	Lemon yellow	134	Blue pigment		Grey	29
		Chrome yellow	53	No. 1	849	Chrome yellow	1
				Blue pigment			
				No. 2	108		
				Violet pigment	30		

wax dispersions that produce smooth flowout and aid in printing ink holdup. Waterproofing agents may also be included if desirable.

The coating loads applied with these formulations range from 3 to 7 lb (20×26—500).

Other formulas for producing dull-surfaced coatings, particularly for printing paper, have been suggested. These formulas usually include some dulling agent. Bradner suggests the addition of uncooked dry starch to the coating mix, about 25% based on the solids, and that this mix be applied in the usual manner and the coated paper calendered. The sheet is passed over a moistening roll and then dried. According to Bradner's disclosure the water is absorbed by the starch grains, which swell slightly and produce a surface that tends to diffuse the reflected light. In a second disclosure, he suggests the use of pulverized cellulose fiber that is claimed to produce a similar effect. Sutermeister states that a fine dull-surfaced paper can be produced by applying a thin wash coat to a precoated and calendered paper. This technique is said to produce a peculiar velvety surface that prints readily, but is easily abraded and marred.

DECALCOMANIA OR TRANSFER PAPERS

The development and utilization of decalcomanias or transfers is an interesting story that dates back to the beginning of the 18th century. In 1826, an Austrian named Rothmüller patented a method of obtaining lithographic transfers, in black and in colors, that could be transferred to tin, wood, waxed cloth, and other difficult surfaces. The early technique was based on a lithographic impression on a sheet of paper previously surfaced with a water-soluble coating, with the order of the printing of the colors for the final effect being reversed. A final coating of white pigment gave the illustration its background. This print was then pressed upon an object that had been previously varnished with copal resin. The paper back was moistened and removed and the picture it supported remained fixed on the new surface.

The technique spread from one European country to another, but Germany was the center of the industry, particularly for the decoration of china and toys. The technique became so popular in the late 19th century that it earned its name, "decalcomania," coined from a Greek root *decal*, meaning off the paper and *mania*, meaning craze. Hand decorations with elaborate styling were costly and took a long time. In contrast the decals, or transfers, offered the advantage of ease of application and flexibility of use and economy. The early volume in this country consisted mainly of stock designs, and these later developed into special designs and name plates. Today decals are used to decorate ceramics, wood, ivory, glass, plastics, and trademark designs, window signs, trucks and mobile equipment. The sizes run from a fraction of a square inch for the state tax seal on cigarette packages, to the thousands of square inches, for trucks and giant air transport planes. In general, they are used to decorate or print anything that cannot be run through a printing press.

The actual manufacture of decalcomanias in the United States was begun by Herman Pfeil in Philadelphia in the early 1890's. The volume grew steadily and the domestic items gradually replaced the imported products. About 1912, the manufacture of ceramic decalcomanias was started. This portion of the industry was an instantaneous success and the volume increased rapidly.

The present-day decalcomania is the result of a highly specialized process. In consists of one or more layers of ink that form a design or pictorial representation built up on a special paper or cloth backing. The "decal" is actually an image of printing ink, lacquer, or synthetic enamel forming a homogeneous film that slips or slides off the moistened paper surface. The paper is a special coated type that allows the built-up film or print to be transferred to the object to which it is to be applied.

The printing or building up of the various films of pigments is generally produced by the the lithographic process, but letterpress, gravure, offset, and silk-screen techniques are also commonly used for special purposes. Some decalcomanias have consisted of as many as 25 to 30 single impressions, although most of them are limited to 3 to 8 individual colors. The pigments used depend upon the requirements, and range from cheap prints, on toys and cheap pottery, to the heat-

and light-stable types, used on good fused ceramics and weather-resistant vehicle decorations. A film of plastic lacquer is often laid down first, the printing is done on top of this film, and a final film of lacquer laid down on the surface. These films are protective in nature, and the final decal is a sandwich structure that can be conveniently handled.

Two main classes of decals, classified according to their method of application, are distinguished. The direct-transfer types are those that are applied to the desired surface, either by moistening a gummed layer on the inside face and attaching the decal to the object, or by applying varnish or shellac to the surface and pasting the face on the object in this manner. In either case the paper back is then sponged with warm water until the paper or cloth backing is released, leaving the print adhering to the object. The "slide off" transfer-type decal is first floated on water, which separates the print from the paper backing. The print is then applied to the desired surface, either by self-adherence or by the use of additional adhesive. The decal print is immersed in water for about 15 seconds and then allowed to stand until the print can be slid off. Too long an immersion washes away too much of the soluble adhesive upon which the adhesion of the decal depends.

A further differentiation in these two classes is whether the print is applied face up or face down. The order of application of the various color prints depends upon which variety is desired and calls for a skillful handling of the colors, as well as a thorough knowledge of the principles of the art work involved. Where the face-down type is used, such as on the back of a window where the print is viewed through the glass, the base or background coat is printed first, followed by the various colored designs that make up the overall design. Where the face-up type which is applied to an opaque background is desired, the various portions of the design that blend to make the final effect are printed in order and the opaque background is printed the last. A special type is the two-sided window decal that can be viewed from either side of the window. One side to be viewed is printed first, each color in the correct order and finally several impressions of opaque background. The design is then repeated in reverse order so that it can be viewed from the opposite side. As a rule, an adhesive is applied over the final print so that the decal can be used as a "slide off" or a direct transfer on the glass.

In printing decalcomanias the register problem is an acute one and air-conditioned pressrooms are generally needed. It is a painstaking, intricate process that requires care and ingenuity to obtain perfect register of brilliant colors. The designs may vary from simple lettering to multicolor pictorial representations whereas the shapes vary from simple squares and circles to complicated irregular contours.

Base-Paper Requirements

The base sheet for use as a backing for the decal may be one of two types. It may be a single sheet to which all coatings and printings are applied, or it may be thin tissue paper laminated to a heavier paper or cloth to eliminate curl and dimensional change and to facilitate the printing and registering operations. In each case, the paper is an absorbent, rather open, unsized sheet, possessing uniform color, formulation, and surface properties and preferably a high wet strength. The basis weights range from 50 to 90 lb (25 × 38—500). The backing sheet is relatively unsized and in some cases may be a waterleaf sheet to facilitate the penetration of water for releasing the decal from the transfer paper.

Preparation and Technology of Transfer Paper

The preparation of the coated transfer paper ready for further coating and printing of the actual decal is the portion of the process that falls in the field of the specialty paper industry. Further treatment and the actual production of the finished decalcomania is beyond the scope of this discussion and will not be covered in any detail.

The first coating laid down on the transfer paper is the coating on the decal side of the sheet and must be water-soluble and ink receptive. This coating acts as a base for the actual printing operations but must dissolve when treated with warm water so as to release the finished decal. The coating is applied to the felt side of the sheet to reduce curl, and must be smooth and uniform and contain no ridges. Brush, roll, or air-blade coating equipment is used to produce the desired results.

The coating may be composed of starch and gum arabic; albumen, glycerin, and starch; egg albumen; starch and dextrose; or it may be any other suitable water-soluble coatings. A fairly heavy coating load is applied because it must produce a smooth printing surface and yet separate the two layers smoothly when it is dissolved out. It is sometimes desirable to lay down a coat of clear or pigmented lacquer over this coating at the converting plant so as to make a better printing base, but this is optional and depends upon the requirements of the customer. General practice is to calender the finished coated paper to produce the required overall smoothness for further printing operations.

This type of sheet is sensitive to atmospheric conditions, particularly humidity, and may curl badly if not stored in moistureproof wrappers and handled in air-conditioned rooms. Curling, of course, not only makes the sheets hard to handle, but also makes good register in the printing press practically impossible to attain. Stewart suggests that the finished transfer paper be coated on the reverse surface with a film of gelatin to exclude atmospheric moisture and to seal in the moisture already present. The coating can be prepared from 10 parts of gelatin and 90 parts of a mixture of water and methyl alcohol in a ratio of 3 to 2. Other water-miscible solvents such as ethyl alcohol, acetone, and tertiary butyl alcohol, can also be used. The coating is applied on a roll applicator and the bath temperature should be maintained at 115°F to obtain the desired solubility and solution viscosity. This coating is claimed to balance the stresses in the sheet and prevent water absorption, both of which lead to curl. It dissolves readily in warm water when the decal is applied to the desired surface.

A technique for preventing curl in the transfer paper and to eliminate registration problems in printing is disclosed by Marksberry. The approach is entirely different from the usual processes for decalcomania preparation and involves printing the desired decal on a waxed or other nonadherent surface, coating the finished print with a water-soluble adhesive, and then causing the transfer paper to adhere to this coating. The printing is thus carried out on a dimensionally stable surface, and the transfer paper acts only as a carrier.

FLOCK-COATED OR VELOUR PAPERS

Flocked or velour papers are specialty coated sheets that simulate the appearance and feel of a pile finish. The actual surface effect may range from plush, velvet, mohair or fur, to a felt or suede. Innumerable types, lengths, and colors of flock are available and the basic varieties are cotton, rayon, wool, silk, and other animal fibers. Cotton flock may be described as having practically no nap. It is not fibrous and produces a dull finish similar to that of suede. Wool flock is somewhat rough and fibrous and is never used where a soft finish is required. It is often used as a durable rug or carpet material.

Rayon flock is probably the most widely used type because the length of the fibers can be closely controlled and can be made to simulate plush, velvet, or velour. The material is available in a number of standard fiber lengths, generally $\frac{1}{32}$, $\frac{1}{16}$, $\frac{1}{10}$ in. and in some cases $\frac{1}{4}$ in. The longer fibers produce softer-textured finishes, but they are much more difficult to cause to adhere satisfactorily to the surface. A wide range of colors is available from the natural shades of gray and tan of the animal fibers and white in cotton and rayon to a full range from red, yellow, green, blue to black and brown in the various dyed materials. The colors can be made washproof and drycleaningproof, as well as completely fast to fading.

Flock coatings become a part of the sheet to which they are applied and their durability depends upon the type of flock, the method of application, and the type of sheet surface upon which they are coated. By printing the adhesive on the base stock in designs rather than as an overall coating, velour prints can be made.

Flock-Coating Technology

Backing Sheet

A flock coating can be laid down on any weight or type of plain, saturated, or treated base paper, film, foil, or cloth. It is purely a surface coating and may cover all of the surface or portions, as desired. If an overall coating is required, the adhesive is uniformly coated over all the surface whereas if only portions must be covered, the adhesive is printed on areas that are to receive the flock. The other areas will not be affected. The backing material should be sufficiently sized or dense enough to hold up the adhesive, and this is the only major requirement. The sheet should possess sufficient strength to meet its use requirements. In general, sufficient flock is used so that the background is completely covered, but to obtain the brightest effects with light-colored flocks, a white backing is advisable.

Flocking Adhesive

The flocking adhesive is one of the most important components of the finished flocked paper. It must be flexible so that it will not crack off when the paper is bent or flexed, it must bond the flock to the paper so that it will not fall out or abrade off, it must be water-resistant so that the flock will not be removed in damp weather, it must age satisfactorily without loss in the properties, and must size the paper and hold up on the surface to bond the sheet and flock together. The adhesive can be used clear, dyed, or pigmented. If it is colored to match the flock, it can increase the depth of shade and uniformity and tend to cover up a poor or defective base sheet. If colored in a contrasting shade, it can be used to produce interesting two-color effects.

Because of the almost dust-like lightness of most flock materials, "stickiness" is a prime characteristic of a good flocking adhesive. A wide range of materials is on the market, and they are formulated depending upon the type of application equipment that is available. Roll, knife, gravure roll, and spray unit application equipment have all been used successfully because the major requirement is that a uniform film is applied to the base sheet.

The adhesives employed fall into three major classes:

1. Water-soluble adhesives.
2. Synthetic resin and rubber solvent base adhesives.
3. Synthetic resin and rubber dispersion type adhesives.

Each has special characteristics and merits, and each is used for particular jobs.

Water-Soluble Adhesives. Water-soluble adhesives formulated from glue, starch, or casein can be satisfactorily used on most sheets, but the finished product does not offer much in regard to waterproofness unless the product is specially treated. Recent developments in this field, however, offer much promise, particularly in water-soluble resins, which are later insolubilized by heat or catalysts. These films tend to be brittle, unless plasticized, and this should also be taken into consideration in their use.

Solvent Adhesives. Solvent adhesives formulated to serve as flock adhesives are available both clear and in a wide range of color. Because they dry rapidly, speed is necessary in applying the flock so that the flock fibers will contact the adhesive-coated surface while it is still wet. Such films, when properly formulated, provide excellent adhesion of the flock to practically any type of surface; cheap or expensive solvents can be used where desirable; the film can be formulated to be flexible and water-resistant; and if the surface to be coated is porous, the adhesive can be compounded so that when coated with a knife applicator the loss of adhesive by absorption is relatively negligible.

Many base resins have been used in formulating such adhesives, but the most popular types seem to be the rubbers or other inexpensive elastomers, as they are soluble in cheap solvents, and do an excellent job in causing the flock to adhere to the surface. Surface skimming is a problem with some of these adhesives. The material should be carefully evaluated from this standpoint, because if the surface skins and a film forms, the flock cannot be piled on, even though the complete film of adhesive has not thoroughly dried.

Water-Dispersion Adhesives. Considerable work has been done on the evaluation of latices for use as flocking adhesives, and although there has not been much commercial application, the experimental results have been promising. The use of solvents is completely eliminated, skinning does not usually occur, and excellent adhesion to most types of surfaces can be obtained. New developments are forthcoming in this field, and many development groups are experimenting with this type of compound.

Flocking Technique and Equipment

The equipment used to apply the flocking adhesive and flock is fairly simple and is basically a coating and dusting machine. A major problem is the possibility of dusting excess fine flock into the various rooms of the plant. If possible, the flocking department should be segregated so that contamination of the rest of the plant is eliminated. Unless the flocking equipment is of an exceptionally foolproof design, it is usually desirable to arrange some type of dust removal system whereby all the air in the flocking department is sucked out through a cyclone collector or a water spray or screening system so that fresh air is continuously being drawn into the plant and any flock-laden air is cleaned and discharged to the outside. The flock can be reclaimed for reuse.

The flock-coating operation is divided into two parts, both of which take place in different portions of the same machine. The first step involves coating or printing the flocking adhesive on the paper, film, or foil. The second involves the actual piling of the flock into the adhesive and then drying out the solvent or water, thus setting the adhesive. The adhesive can be laid down on the surface by a roll, reverse roll, knife, gravure roll, or spray technique, depending upon the type and formulation of the adhesive and the surface to which it is being applied. The amount of adhesive laid down varies with the flock type used and the porosity and smoothness of the surface of the sheet. Two general rules can be given: (1) As the length of the fiber flock is increased, so must the thickness of the adhesive film be increased and (2) on applying ordinary flock fibers (1/32—1/10 in.) as the thickness of the adhesive film is increased, so will the thickness of the resulting flock mat be increased.

The adhesive-coated paper is then passed into the flocking unit.

This is usually an enclosed chamber, preferably run under a slight vacuum so that excess flock is carried out and into a separate collection chamber where it is recovered for reuse. The flock is generally held in a hopper that continuously feeds it on the adhesive-coated surface by means of a vibrating screen or a special nozzle, the main purpose of which is to distribute the flock fibers uniformly over the surface area. The flock should be distributed while the surface is still wet because the ends of the fibers then sink deeply into it and become permanently anchored when the adhesive dries. If the adhesive is allowed to dry partially and to become tacky, only the fibers get a surface hold.

At the same time that the flock fibers are falling on the wet surface, agitating and orienting means will pile the fibers deeply and uniformly into the adhesive. This is usually accomplished by vibration or electrostatic orientation. With the vibration technique, the coated surface is agitated by a series of beaters whose vibrating bars strike a flexible screen surface positioned just under the moving adhesive-coated web. As the bars rotate at high speed, the screen surface is set into motion, creating static and causing the flock fibers to bounce 1 to 2 in. above the screen. An electrostatic unit sets up an electrical field that causes an orientation of the fibers and improves their packing. Air jets have also been suggested to improve the flocking operation. The basic purpose of the operation is to have all the individual fibers adhered at one end piled into the adhesive, as well as to have sufficient flock present on the sheet for coverage. The vibration method of flocking usually produces excellent results as it packs the fibers tightly into the adhesive—as many as 200,000 per square inch.

The flock-coated sheet can then be passed through a drying chamber where the adhesive film is dried and set. Hot air, steam coils, or infrared heating units have been used satisfactorily. The sheet may be then passed under a vacuum nozzle or a rotating brush to remove nonadhered flock and brush up the surface The finished sheet can then be rewound into rolls, slit, or sheeted as desired, or printed in further converting operations.

Applications

Flock coated paper has many applications. Box and fancy packaging papers, specialties, display papers, as well as cover stocks are all being produced. This paper is also used in curtains and drape stock; greeting cards; menu, book, brochure, and memo covers; drawer, cabinet, and box liners; lamp bottoms; gasket stocks; floor covering in automobile rear decks; interior walls, and countless other purposes.

Marbled Papers

R. S. BRACEWELL

The process of marbling paper is an old and specialized art developed in the early 17th century and probably existing at a much earlier date. It was first practiced as a hand process for producing certain designs by making patterns in floating colors on a mucilaginous liquid or ground and then transferring these to a sheet of paper or the smoothly cut edges of a book by a dipping technique.

The early hand-marbling processes were based upon the use of a ground mucilage prepared from gum tragacanth, Irish or carrogeen moss, or another hydrophyllic colloid. In a typical formula 1 lb of gum tragacanth was soaked in 2 gal of water for a week, and as it slowly dissolved, additional water was added until the desired consistency was obtained. The final mucilage was strained through a fine cloth sieve to insure the elimination of any lumps or dissolved matter. The colors, which consisted only of pigments, were prepared for use by grinding them with a beeswax composition in the proportion of 1 oz of wax preparation to 1 lb of color pigment. The beeswax composition was prepared by melting together 2 lb of beeswax with $\frac{1}{4}$ lb of soap and adding sufficient water to make a frothy mass of curds that could be crumbled between the fingers and be ground with the color without sticking to the milling stone.

For preparing the marbling composition, the compounded pigments were mixed and thinned with water and ox gall. In some cases, in the preparation of special patterns, olive oil, linseed oil, turpentine, kerosene, or a solution of alum was utilized. This color was dropped or spattered on the surface of the ground mucilage or substrate. The wax acted to bind the pigment to the paper and permitted a later glazing operation. The purpose of the ox gall, alum, and various oils was to vary the surface tension of the mucilage and to permit the formation of the patterns.

The heavy mucilaginous consistency of the ground gum and the wax mixed with the pigments resulted in a tendency of the colors to float on the surface. The colors desired were spattered or dropped on the surface of the ground and the pattern of swirls, ribbons, or spots was developed by means of a comb or other instrument. The sheet of paper was then rolled across the surface and the pattern transferred to it. The marbled sheets were then hung up to dry on frames. This process was slow and laborious and the art was a closely guarded secret and protected and passed down by the apprenticeship system. Much of the marbled paper was later sized or flint-glazed to produce a desirable glossy surface.

The development of this art from a sheet to a web-processing operation was carried out by the Marvellum Company. Careful experimentation showed that pigment colors and metallic pigments could be floated on the surface of a water bath and transferred to a continuously moving sheet of paper and thus handled in a web form. The early experiments with oil as a binder and various oily thinners to cause spreading were carried out by George Sensenay. The designs could be made with combs or other forming tools and patterns of different character from the older type of marbling were produced. This field was explored both by changing the formulation and by different machine applying techniques.

Base-Sheet Requirements

The requirements of the paper for continuous marbling are a minimum absorption of water previous to its being blown off the surface by a slotted air nozzle and sufficient sizing so that the oil binder is not absorbed and does not leave the pigment or the metallic setting on the surface without proper bond to the paper. The surface of the paper has to be smooth enough so that the pattern retains its details and at the same time has enough tooth for handling to prevent offset and smearing. Pigment-coated papers are not suitable for this marbling. Mica-coated papers are used quite successfully.

Marbling Technology

Spot Marbling

This process was first carried out by means of combs and other forming devices. To produce marble patterns it was necessary to work on the surface of water that was comparatively free from flow. This was a simple matter when done by hand in a pan of water, subsequently rolling the sheet along the surface and picking up the pattern. The problem was to produce the same results with a continuous web.

A circular machine or "merry-go-round" that could handle paper up to 30 in. wide at a speed of approximately 50 ft/min. was built. The circular pan with a wooden fabricated base, 36 in. wide, was mounted on a circular monorail track, approximately 25 ft in diameter. A gear rack on the bottom outside of the pan base facilitated the drive of the pan containing water of about 1 in. depth. When the drive was started, there was of course a tendency for the water to lag, but this soon became a constant and a negligible factor.

Opposite the pan drive was a set of contact rolls to bring the web of paper down to the surface of the water and to pick up the pattern. After blowing off the excess water by means of a jet of air emitted from a slotted pipe, the paper was carried over drum driers. The half of the machine on the approach to the contact rolls was in general used for successively applying the colors and forming the pattern. The half of the machine following the contact rolls was used for clearing the water surface, if necessary, before the applications of the colors.

Four color stands that straddled the moving tank were set up. At first, special cylindrical brushes were used to spatter the color on the water. These consisted of a wooden roll on which were spaced, in approximately a 2 in. square cross section pattern, little tufts of bristles mounted on a wire stem about 4 in. long and hinged on a staple on the roll. As this special brush slowly revolved, the tufts fell down and drew across the surface of a color roll picking up its charge of color. As the brush left the color roll it swung on the hinge on which a stop was arranged. This was just enough to throw the color from the tufts on the surface of the water in drops sufficiently large to form the desired pattern. These special brushes did an excellent job but their main drawback was their cleaning and the wash up. The brushes were later replaced by slowly spinning disks that did satisfactory work. These allowed the color roll to be cut down to a 2-in. face from 30 in. and thus simplified the wash up and allowed great economy in color.

As previously stated, stationary or moving combs were used for variation of patterns. Different effects could be obtained by varying the oil-vehicle formulation. For instance, a varnish of high gum content produced a characteristic pattern under the action of the comb, whereas a formulation based on nonblushing lacquers produced a different effect.

This machine was later widened to marble 36-in. webs. This was about the limit for the diameter of machine because the difference in the circumference of the outside and the inside of the pan caused some distortion of the pattern on the web.

Grain Marbling

In contrast to the spot marbling carried out on a moving though nonflowing water bath, grain marbling was produced on a water surface flowing down a stationary inclined tank. This tank was straddled by a single stand handling four colors that were led down pipes to an inclined baffle board that made contact with the surface of the water over the entire width of the tank. These four pipes were in the form of a pendulum that swung back and forth across the tank depositing a blend of the colors on the baffle board. This blend of colors filmed out on the water, was formed into a pattern and was then picked up by the web and handled just as on the other machine.

A so-called rippled effect was produced on this machine by setting a dam with an apron across the tank just ahead of the contact rolls. As the water carrying the colors on the surface flowed down the apron a ripple was formed, and by making contact at this position, rippled patterns were obtained on the web.

Applications

Marbled papers were used as box papers, envelope linings, gift wraps, cover paper, photomounts, and calendar mounts. For box papers, envelope liners and gift wraps, much use was made of gold and aluminum as well as of the brilliant colored metallics. In the cover paper and photomount field, marbling was done in colors with perhaps just a vein of metallic pigment running through the design. Grain marbling was suitable for this purpose and was perhaps more popular than spot marbling.

Marbling was also used in laminating base papers for panels and table tops. Later this was replaced by printing, which was a cheaper process, but unusual effects were still obtained by marbling.

Marbling was developed for top-coating rubber and similar films, but later marbled effects were achieved by blending colored plastics in the film casting or coating operation.

Filled cloth that was marbled for book binding was handled quite successfully. Some attempts at marbling silk and cotton were made and very attractive samples were produced, but the trade did not take hold of these developments. A method of marbling vat colors on cotton was developed in which the vats were marbled on the cloth in pigment form; the dye was subsequently developed on the fibers. A method was also developed in connection with one of the large textile concerns wherein a water-soluble dye was marbled directly on fabric but neither of these became of commercial importance.

Some indication that a demand would develop for the return of the spot marbling in the fancy paper field has been noted but with the modern treatment of fancy papers it has been possible to offer items that have satisfied all the tastes and could be made more economically than by the marbling process. The wood grain has remained popular and will probably continue because it has an interesting natural appearance.

BIBLIOGRAPHY

N. McClelland, *The Practical Book of Decorative Wall Treatment*, J. B. Lippincott, New York (1926).

P. Ackerman, *Wallpaper: Its History, Design and Use*, F. A. Stokes, New York (1923).

C. W. Ward, *Wallpaper: Origin and Manufacture*, Putnam, New York (1921).

H. Heinrich, "Requirements for Wallpaper Pigments," *Farbe und Lack*. **43-4**, 97-8 (1932).

Ducrou, "Starch in Wallpaper," *Recherches et Inventions*, **16**, 254-7 (1935).

M. Weilerswist, "*Manufacture of Wallpaper in Historical and Modern Times, Papier-Ztg.* **64**, (35) 1008-10 (1939).

W. J. H. Hinrichs, German Patent 610,018, Preserving Wallpaper by Application of Rubber Latex Film (1935).

C. J. Wernlund, U. S. Patent 2,073,666, Hardening Proteins by Means of Insoluble Aluminum Compound and Acid-Forming Ammonium Salt (1937).

J. W. Close, U. S. Patent 2,284,800, Washable Wallpaper Using Corn Protein, Alkaline Resinate and Bleaching Compound (1942).

F. K. Schoenfeld, Canadian Patent 416,754, Wallpaper Coated with Plasticized Vinyl Chloride Film (1943).

National Starch & Chemical Co, The Birge Company, Inc., U. S. Patent 3,122,446, Process for Making Remoistenable Adhesives (1961).

William Katzenbach, *Practical Book of American Wallpaper*, Lippincott, Philadephia (1951).

W. M. Bain, A. W. Neubauer, and R. A. Olson, "Synthetic & Protein Adhesives for Paper Coating *TAPPI Monograph* No. 9 (1952).

E. A. Barclay and A. N. Thompson, Preparation of Paper Coating Colors *TAPPI Monograph* No. 11 (1953).

R. L. Davidson and M. Sittig, *Water Soluble Resins*, Reinhold, New York (1962).

chapter 5

Functional Papers

Miscellaneous Applications

R. H. MOSHER

Many decorative papers do not fit into the overall pattern of discussion, and there are also functional papers that cannot be classified directly into any other group. Such papers are produced because of some functional property separate from any decorative value that the sheet may possess. For example, a fancy boxpaper may be attractive to look at, but it may also possess antitarnish properties.

Such papers are: (1) coated chart papers; (2) safety and anticounterfeit papers; (3) magnetic recording paper or tape; (4) antitarnish papers; (5) electrically conducting papers; (6) mold, bacteria, insect, and rodent resistant and preventative papers; (7) flame-retardant papers; (8) plasticized papers.

These sheets are often produced merely for their functional characteristics, yet many of these characteristics can be incorporated into regular decorative papers.

COATED CHART PAPERS

Coated chart papers are made for use with recording instruments where a metal stylus acts as the recording head. The coated surface shows a permanent black marking wherever a metal surface is drawn across it. Such a paper, in a recording instrument, will show a continuous or intermittent record without the need for an inking pen or pencil. When it is desirable to make an additional notation or marking on the chart paper, this can be made with a regular pen or pencil, or merely by using the edge of a coin, a key, or other blunt-edged metallic object.

The pigment in the coating used on such papers is generally a baryte or blanc fixe or some combination of the two. Clay or other extender can be used but it is seldom applied. Casein is the usual adhesive. A relatively high coating load, about 10-12 lb (20×26—500) is applied to the paper, as a light touch of a stylus must definitely mark the surface.

The coating can be applied by a brush, roll, air-knife, or spray coater and the only problem is the possible settling of the heavy pigment. Good agitation will overcome this trouble. The coating dries readily and the finished sheet should be given a light calender finish to smooth the surface, but not sufficient to give it any polish.

A typical formula for coating such papers is as follows:

Blanc fixe pulp (66% solids) 500 lb
Casein 57.5 lb
Water 37.5 gal

Sufficient cutting agent for the casein, eveners, and foam killers, is also added to make satisfactory mix.

SAFETY AND ANTICOUNTERFEIT PAPERS

Safety papers are special papers treated to make visible any attempts to alter the writing or printing on them. They can be produced by the addition of certain chemicals in the beater when the paper is being made, by a surface treatment or coating, or by printing or embossing an over-all design on the surface with sensitive inks. Methods of introducing the compounds in the beater are of no significance to the converting industry, but coating and printing techniques will be discussed. Anticounterfeit papers are papers that have been made or treated so that their identity can be verified by some easy visual or chemical means to prevent counterfeiting.

SAFETY PAPERS

The principle upon which all safety papers is based is their ability to exhibit telltale reactions when they are altered or tampered with. Some inks or coatings change color when acids, alkalis, or bleaching agents are used; some blur and diffuse the coating or printing over the area in contact with the agent used; and some show bleeding of a water- or oil-soluble dyestuff when any attempts are made to remove or alter a stamp or seal. The basic feature is similar whether an over-all clear or colored coating or an over-all printed design is used, therefore, both types will be discussed without specific differentiation.

By using an ink or coating, the color of which is a function of the acid concentration or pH, any employment of acid bleaching agents or alkaline compounds can be detected from a variation in the shade of the design or surface. Inks can be made acid-sensitive by the addition of zinc chloride or hexamethyl and pentamethyl monoethyl-*p*-rosaniline, the diazo dye from tetrazotized *o*-toluidine, sodium naphthinate, sodium hydroxide, and monosodium phosphate. Compounds that contain metallic salts of fatty acids change tint when treated with a detergent. Aniline hydrochloride discolors by the decomposition of the hydrochloride when a bleach comes into contact with the paper. Many common dyestuffs change color with change in pH.

Colorless or slightly colored manganous compounds that are changed to colored manganic compounds by oxidation have been suggested as wash coatings. Dioxy- and trioxy-benzoic acids and their derivatives give color reactions with even weak oxidizing agents and these are almost impossible to reverse or bleach out by application of the most powerful reducing agents. The use of a colorless pyrogallol wash coating that turns brown on introduction of an oxidizing bleaching agent has been suggested. A paper containing polyoxybenzoic acid does not react with acids, but does provide an almost complete protection against partial chemical falsification or erasure by color change. A special preparation is disclosed by Baush and Schroth that is sensitive to oxidizing agents; the resulting stain cannot be eliminated by reducing agents. The sheet is treated with a wash coating of ferric salts, such as ferric acetate, and then a suitable acid-soluble ferrocyanide, such as lead ferrocyanide. The final solution used is a mixture of equal parts of pyrogallol carboxylic acid and pentagalloyl glucose brought to a pH of 3.0 to 3.5 by the addition of alkali.

These authors also disclose the use of another class of reducible inorganic salts that are resistant to decomposition by light and that form differently colored substances upon decomposition by oxidizing agents. The salts used are selected from a group consisting of manganous tellurite, barium tellurite, and alkali selenite.

To guard against the sending of secret messages by war prisoners who might be writing with invisible or sympathetic inks, papers have been produced that immediately develop writing with invisible inks into visible characters without the application of heat or chemical reagents. The sensitive coated paper first utilized in these experiments consisted essentially of a paper base, coated with clay, containing a mechanically incorporated dyestuff that reacted when moistened or when written upon with an acid invisible ink to produce a green color.

The paper did not react, however, with alkaline invisible inks to produce distinctly visible writing. Therefore, a second ingredient was incorporated into the paper coating. This ingredient reacted with alkaline sympathetic or invisible ink to produce a red coloration. The paper thus manufactured immediately showed up the writing with acid or neutral invisible inks in green color; with alkaline invisible inks the writing appeared in red. In this way, all markings on the paper were visible to the censors or other authorities.

In evaluating the permanence of these developing colors, surveillance tests have shown that they will last for months and surely a sufficient length of time to be clearly detected by censors. The advantage of this device lies in the fact that a prisoner of war may be permitted to write freely upon this specially prepared so-called "Sensicoat" paper with any kind of visible ink but cannot use any invisible writing inks because all of them—acid, neutral or alkaline or even plain water itself—develop color when brought into contact with the paper.

The paper as originally developed was a 56-lb coated sheet, basis 17×22 in. per 1000 sheets, with the color sensitizer incorporated into the coating. Its high cost and heavy weight were factors influencing the development of a lighter, uncoated, and much less expensive paper.

The requirements for such a paper were severe. It was to be an uncoated stock, very sensitive to writing with all types of solutions that could be used as sympathetic inks, as unreactive as possible to humidity and ordinary handling, but nevertheless capable of being manufactured at a substantially lower cost than the first paper used for this purpose. Intensive research revealed ways of meeting these requirements.

A series of laboratory and plant experiments was conducted, ending in a completely satisfactory paper that could be made in one plant from start to finish, including the grinding of the special dye in the sensitizing formula. The new paper was called "Anilith" to differentiate it from the original "Sensicoat."

These examples are representative of the type of safety papers made to resist alteration of printing or writing of checks, legal documents, stamps, and other security papers and to show up attempts to use invisible inks. Another type of paper is used in the preparation of government codes, military orders, and other documents that must be preserved in secrecy and that must be destroyed in an emergency so as to prevent their falling into the hands of the enemy. Some such papers are made so that they will burn and leave practically no ash, but in many cases this is not a sufficiently fast and reliable procedure for destruction, telltale smoke may be given off, and unless the ash is completely decomposed, it may be possible to reconstruct the contents by means of recently developed scientific detection techniques.

According to Foote and Guertin a paper can be coated with a special composition that can be destroyed by merely submerging or dipping the sheet in water. The paper and coating are so constructed that the water permeates almost instantly and causes a chemical reaction by which gas is developed in large volume and in the form of bubbles that loosen the coating from the paper. At the same time, the bubbles in the material of the coating destroy it. Any matter printed or written on the coating is obliterated and destroyed to such an extent that nothing remains to give any information about what appeared on its surface before destruction. A waterleaf sheet is recommended, and this may be surface-sized with starch or not, as required. A suggested coating for the printing side of the paper is as follows:

	Parts
Methyl cellulose	6.5
Calcium carbonate	0.78
Water	sufficient to make a coatable composition

This coating can be applied by a normal coating procedure and will be decomposed at a high hydrogen ion concentration. Small amounts of plasticizer may be added if desirable to produce additional flexibility, but are not necessary for good results. Instead of methyl cellulose, such binders as gum arabic, dextrine, or glue can be used.

When this coating has been completed, a second coating is laid down on the back side of the sheet. This coating is applied at as high a viscosity as possible to prevent its penetration into the sheet and is composed of methyl cellulose or another binder and an acid such as tartaric, citric, maleic, or lactic. Plasticizers can also be used. The acid is dissolved in an anhydrous solvent such as anhydrous alcohol so that a reaction will not occur during the coating operation.

ANTICOUNTERFEIT PAPERS

According to Kantrowitz several interesting developments in the field of anticounterfeit papers were used extensively during World War II. Papers that had been made as identifiable types, specifically for currency or bank note paper, had been based on the use of short red and blue silk fibers that were incorporated into the sheet when it was manufactured on the paper machine.

These silk fibers are visible to the naked eye because they differ from the cellulosic fibers from which the paper is fabricated. Identifiable paper was also proposed in which the mark of identification comprised fibers that fluoresce under ultraviolet light.

In the printing of war ration books that precaution was taken to discourage the counterfeiting of stamps by incorporating into the paper structure fluorescent cellulosic fibers invisible in ordinary light but visible under the influence of ultraviolet light.

In connection with the printing of subsequent ration books, the technical division of the Government Printing Office engaged in research to develop identifiable paper that would distinguish genuine from counterfeit ration coupons by means other than the use of fluorescent fibers.

It was found that ferric chloride-treated fibers incorporated into the structure of the paper to the extent of 0.5% would remain colorless and invisible in the sheet until treated with certain reagents, such as potassium ferrocyanide and orthophosphoric acid. They then become visible and individually identifiable by acquiring a distinctive and permanent blue color.

These fibers presented several advantages not obtainable with silk or fluorescent fibers. One advantage is that they are invisible within the paper structure and do not reveal their presence to counterfeiters. Another advantage is that they are basically identical in composition with the untreated fibers of the entire sheet of paper.

As compared with fluorescent fibers, paper containing chemically treated fibers can be used in the field without necessitating expensive ultraviolet light equipment for tests. Another advantage of chemically treated fibers is that unlike fluorescent fibers they do not lose their effectiveness upon exposure to sunlight, bleaching agents, acids, alkalies, water, oil, or gasoline and, therefore, remain in the paper structure as a permanent means of identification.

As a result of the passage of the Soldier Ballot Bill by Congress, it became necessary to supply the soldiers with forms that could be readily and positively identified as genuine. The identifying characteristics of this paper were to be such as could be incorporated very quickly and with existing paper-making equipment. They were to be able to withstand the action of salt water, heat and high humidity, and to require no complicated tests to reveal their presence.

Several samples were submitted to Army and Navy officials and upon their approval there was included in the paper a safety feature that consisted of printing on both sides of the ballot a design with an ink invisible under daylight or artificial light but which fluoresced brightly when stimulated by ultraviolet radiations. The fluorescing design was composed of the phrase, "Official Federal War Ballot," horizontally repeated and separated by a star between each repetition.

The work of the Government Printing Office as outlined by Kantrowitz shows how far the development of such specialized papers can be carried when specific problems must be solved.

The anticounterfeit papers can be made by the use of such identifiable printing or coating techniques but they are also produced by laminating together a number of plies of colored paper so that when a document or ticket is torn in half, the arrangement of the colored layers identifies its authenticity.

MAGNETIC RECORDING PAPER OR TAPE

An interesting development is a metallic-coated paper that serves to supplement or replace wire as a medium for recording sound. Such paper is coated with a magnetically reactive metal powder, which acts in the same manner as the wire in becoming activated in a magnetic field and recording and reproducing sound effects.

In the process of sound recording by this technique, the sound waves are picked up by a microphone and are converted into corresponding electrical energy. This is amplified electronically and fed into a special magnetic tape recording head (Fig. 5-1) that converts electrical energy into corresponding magnetic fields of varying intensity. Magnetic tape moving at a constant speed is brought into direct contact with or very close to the recording head. Magnetic patterns are formed upon the tape that correspond in wave length, intensity, and polarity to the acoustic pattern of the original sound waves. When the tape is played back, the magnetic tape is passed at a constant speed and in direct contact with or close to the pickup head magnetic tape. The magnetic flux on the tape passes through the pickup head and creates electrical impulses subsequently amplified and fed to a loud speaker that converts the electrical energy back into sound waves corresponding to the original sounds picked up on the microphone. An alternate playback system passes a constant-intensity magnetic field or bias voltage through the tape which is altered by the flux pre-

Fig. 5-1. Details of recording and erasing heads on playback instrument that uses paper tape. (*Courtesy of Amplifier Corp. of America*).

sent in the tape itself; this variation creates the necessary impulse for amplification of the original sound.

Paper tapes possess several advantages as compared to wire. The tape hangs limp, and will not coil, knot, and break as does wire. It can be easily edited and spliced with pressure-sensitive tape. The tape can be made in any widths suitable for the machine and a number of different sound tracks can be run in parallel. At the present time the cost differential is favorable but if the cost of the wire can be reduced, this advantage may be eliminated.

The actual medium upon which the magnetic modulations are impressed is a layer of finely powdered crystalline magnetic particles uniformly coated on a paper base. Magnetite and purified iron oxide as well as a number of special alloys have been used. High physical strength and the minimum possible caliper or thickness are essential for the paper base. The density of the sheet must be absolutely uniform and the surface must be as smooth as possible to obtain the best results. Some of the paper now in use is about 1.5 mils in thickness and an even thinner medium is desirable. The metallic coating must be bound firmly to the paper surface and the film must be flexible and absolutely uniform. Synthetic resin adhesives are generally used to bind the pigment to the paper surface. The denser the coating the better the fidelity in the recording. The magnetic coated layer is being run 0.5 mil in thickness by one producer, the thickness of the coating has been found to be quite important in regard to the effectiveness of the response.

A recent investigation in magnetic tape design also closely relates the high-frequency response of the tape to its thickness. The thinner the tape the better its high-frequency response. The residual magnetism retained by the tape (all other factors remaining constant) is also a function of the thickness. The thicker the tape the more magnetism it retains, and this causes overtones and noise during the playback. The present tape made at 2 mils overall thickness seems to be a reasonable compromise between tape strength, effectiveness of response, and undesirable magnetic phenomena.

The recorded wave length of a given frequency is a direct function of the speed. The lower the tape speed the shorter the wavelength. As very short wavelengths correspond to very short bar magnets, and as extremely short magnets exhibit a marked self-demagnetization property, it is desirable to run the tape at sufficient speed so that half the recorded wavelength is not much smaller than the recording head gap, and is at least ten times larger than the average magnetic particle size. With proper design, a magnetic recorder utilizing a tape speed of 7½ in./sec can record and reproduce a 9000-cycle tone. Magnetic wire requires a lineal speed of 24 in./sec to reproduce a 5000-cycle tone.

The same magnetic tape can be used many times as the magnetic patterns impressed upon the tape can be readily erased electronically without affecting the tape in any way. The tape merely needs to be demagnetized and it is ready for a new sound track to be imposed upon it. Such a demagnetizing process consists in subjecting the tape to an intense supersonic magnetic field of gradually diminishing intensity so that as any given previously magnetized particle leaves the demagnetizing field, it resumes its original unoriented state. In wire recording another problem may arise from the fact that an uninsulated wire is wound in direct contact with other magnetized wires. As a result, magnetic cross interaction can take place and a cross transfer of signals can occur. This phenomenon is virtually eliminated when paper tape is used, as the paper tape backing acts as an insulating layer between two magnetic layers in the spool.

One type of tape now on the market has the following specifications:

Composition	Metallic powder coated on paper tape
Tape dimensions	Width—¼ in.
	Thickness—0.002 in.
	Coating—0.0005 in. (thickness)
Break load	6 lb
Dimensional stability	Excellent

The machine designed to operate on this tape has the same head for the recording and playback circuits. It has a gap of 0.0005 in. and the tape runs across the top of the gap and not through it. In recording, a bias signal of 18-20 kilocycles is applied to 130 turns of the winding and the audio signal to the remaining 30 turns. For the playback, the signal is picked up from the 130-turn winding. This tape is claimed to operate satisfactorily for recording up to 6000 cycles/sec with a speed of 8 in./sec. Further development may permit a wider frequency range for fidelity or a lower tape speed for economy.

ANTITARNISH PAPERS

The antitarnish papers in general use are usually based upon 10- to 16-lb tissue where the sheet is to be used directly for the wrapping of individual pieces of silver or metal parts. Antitarnish requirements are also important for special coated and printed papers to be used in box coverings, decorative panels, napkins, labels, household wrapping papers, and other applications where the paper may come into direct contact with polished metal surfaces. Silver antitarnish paper is white and usually made from an all-rag or a rag-and-sulfite fiber base. Other antitarnish metal wrappings are produced from 100% kraft and are brown in color. Any pulp used in the production of such papers must be washed sufficiently to reduce the sulfur content to less than 0.005%, and the paper must be dense and free from pinholes.

According to Rowe and Kress, there are two major lines of thought as to what constitutes a good antitarnish paper. One group of investigators has claimed that the tarnishing is caused by the paper itself. In the case of silver, tarnishing is caused by sulfur compounds in the paper, whereas in the case of steel and aluminum the trouble results from the presence of free acids. Paper for wrapping silverware should be free from sulfur, sulfides, and dyestuffs containing sulfur. Some researchers feel that the paper should not only be free from harmful substances, but should be dense enough to prevent the passage of tarnishing gases.

The nature and origin of substances causing the tarnishing of metals wrapped in paper has caused much dispute. As stated, wrapping paper should be free from acids, chlorine, and sulfur compounds. One author claims that very few papers, when used, are completely free from tarnishing ingredients. Some state that it is dangerous to guarantee sulfur-, chlorine-, or acid-free papers as they may not remain free of these substances all throughout storage and transit.

A second group of investigators believes that tarnishing is due to conditions outside of the paper and that hydrogen sulfide and other sulfuric gases are the main trouble-producing factors with silver, whereas carbon dioxide and oxygen are conducive to tarnishing of steel and aluminum. This school of thought maintains that a dense and impermeable paper that is waxed or coated or a sheet that contains chemicals capable of adsorbing these gases before they reach the metal is the best solution to the problem. Most authorities today agree that the major source of trouble, particularly with silver, is the presence of sulfides, and the paper must not only be free from them, but some method of barring them from the metal surface is required.

Much work has been done on various coatings and impregnants for use in the manufacture of antitarnish paper. A mixture of light and heavy oils for impregnating has been suggested. A treatment with a soluble salt such as zinc acetate is claimed to be sstisfactory. Other suggested salt treatments involve the use of lead acetate, cadmium acetate, copper acetate, copper sulfate, nickel acetate, and copper oxide. In all these cases the principle of protection involves a preferential absorption of the hydrogen sulfide gas by the salts rather than by the silver piece contained in the salt-impregnated wrap. Such treatments have been found to be quite satisfactory and probably represent the best available protection at the present time.

Other suggested protective methods have involved such soluble salts as those listed above together with waxing, the use of copper sulfate and copper oxide added in the beater, and the treatment of the paper with alkaline earth bicarbonates.

ELECTRICALLY CONDUCTING PAPERS

Papers that allow the passage of electric currents and simultaneously record the passage of these currents have a number of applications, e.g. the facsimile recorder, electronic tuning devices, and the Marconi depth sounder recorder. Others are represented by the various conducting and antistatic treated sheets. In most of the first type, signals of varying strength are recorded on paper by means of a moving stylus. The current flows from the stylus through the paper and to a metal plate that supports the paper. These fall into two categories, those in which the stylus takes part in the reaction and those where an inert stylus is used. In these papers, the current must produce a sharply defined stain along the path of the stylus, the stain should develop instantaneously, show an increase in intensity as the voltage is increased, and should not fade. The principal objection to many of such papers is the instability of the trace and the impossibility of storing them for long periods without drying them out. Such drying must be followed by rewetting before use. The question of mold growth is important in this respect, and must be considered, particularly in the tropics and during the summer season.

The second classification includes those sheets that can be surface plated or receive a metallic deposit from solution by electrolytic means and nonconducting papers or films that behave unsatisfactorily because of the buildup of surface static. The conductivity of plating or conducting papers depends, to a great extent, upon the voltage used and this may be a limiting factor on the system. The antistatic treatment varies with the conditions under which the system will operate, i.e. the humidity and presence of conducting bodies.

Conducting and Recorder Papers

These papers have been made by means of several methods including treatment with various salt solutions and the use of starchiodide, as well as coating and saturating with carbon black, acetylene black, or graphite. Coatings and treatments with aluminum, copper, and other metal powders and mixtures of the metallic powders and glycol stearates have been utilized. The pigments may also be beater-dispersed.

Examples of salts used to saturate recording papers are benzidine and sodium bromide; pyrogallol, monosodium phosphate, and sodium nitrate; p-anisidine and sodium bromide; and o-toluidine and sodium bromide. Such papers when tested with an irridium stylus passing over the paper at 0.2 ft/sec or 0.5 ft/sec, resting on a metal plate with an applied voltage of 0 to 7.5 V, gave satisfactory discoloration. The efficiency and stability of such papers vary widely and their usefulness depends upon the specific application.

Carbon-containing conducting papers can be made by beater dispersion of the conducting medium, or by coating or saturating the sheet with the compound. Acetylene black and graphite are the best for this application. The conductivity of paper containing such blacks depends greatly on the resistance built up between the particles, and the binder, if any, should be carefully chosen. Metallic stearates are often used. The conductance of a beater-dispersed or saturated sheet depends to a great extent upon the amount of conducting material present, and a choice must often be made between strength in the sheet and the desired conductance. Such papers are generally black in color—a disadvantage because the paper cannot be used for recording.

The metallic-base powders are often used to produce conducting papers but their electrical conductivity is often found to be impaired by the presence of oxide films on the particles, which act as insulators. The addition of stearic acid esters of diethylene glycol, monohydric alcohols, simple polyhydric alcohols, sugars, hydroxy carboxylic acids, aldehydes, and ketones are all recommended. These substances produce a film readily and possess a low melting point that simplifies the calendering operation. Where color is desired so that the sheet can be used in recording or for facsimile, lead thiosulfate, lead thiocyanate, or mercuric sulfide may be added when the metallic portion is coated on or combined with the sheet.

In some cases it is desirable to combine a conducting paper and a recording surface, especially for fine-definition facsimile or electrostatic recording work. This is particularly true where fine half-tone pictures, charts, and diagrams must be reproduced. The body of the blank may be made conducting by impregnation with such salts as sodium, potassium, and aluminum iodides, sulfocyanides, bromides, chlorides, nitrates and sulfates. The recommended carriers are diethylene, triethylene propylene, butylene, or polyethylene glycols. As a coating on the surface that will take the marking, the water-insoluble sulfur-bearing compounds of metals are suggested. These compounds comprise thiooxalates, thiolates or mercaptides, thioacetates or thioglycolates, and thiocarbamides of metals such as copper, silver, mercury, and lead. Such compounds are white, and turn black where the current passes through because of reduction or conversion of the compound. The sulfides produced are insoluble and permanent in color.

The base paper used with such facsimile recording units varies with the desired results from a cheap groundwood-containing sheet to a multiple-coated, smooth-surfaced paper for accurate reproduction.

One recommended paper was a 60% sulfite—40% bleached kraft sheet of high wet strength. The sheet was treated with p-chloro-m-cresol, 0.05%; trichlorophenol, 0.05%; or sodium fluoride, 0.5% as a fungicidal agent. After treatment, saturation, or coating, the paper was calendered to make a smooth recording surface.

Conducting and Antistatic Papers

An antistatic problem refers to the accumulation of electrostatic charges on the surface of nonconductors such as sheets or rolls of some plastic films and coated papers. The usual procedure involves coating the surface with an extremely thin layer of a conducting medium such as a polyhydric alcohol, sugar, metallic stearate, amino salt, or metallic salt film. Such films are extremely effective and do an excellent job of bleeding off any electrostatic charges as desired.

MOLD, BACTERIA, INSECT, AND RODENT RESISTANT AND PREVENTATIVE PAPERS

Packaging papers of all types, but particularly food packaging papers, are susceptible to attack by molds or fungi, bacteria, and insects, and even to some extent by rodents. Each of these types of attack are rather specific, both from the standpoint of the result if unchecked and the method of protection. The package will be discussed here as that is where the property is of importance.

Mold and Bacteria Resistance

The mold- and bacteria-resistant papers are designed to resist the growth of mold on the packaging medium; the mold preventative papers are mold resistant and in addition act to inhibit any growth on the material contained within the package. The growth of molds generally produces an objectionable discoloration and blotchiness on the surface of the package or product that often gives off objectionable musty odors. Fungi may also produce considerable quantities of organic acids that may react with either the coating or saturant in the packaging medium or with the product itself. The presence of molds or fungi also brings about an unusual concentration of moisture at the infected area. The growth of molds and fungi may occur on the coating or saturant or it may occur on the base paper or film itself. Protective compounds may be incorporated into the paper, the saturant or the coating, and all have been used.

Insect Resistance

The insect repellency can also be divided into two parts: the repelling of insects from the exterior of the package and the erection of a barrier that they cannot penetrate, and the prevention of the growth and development of insects from eggs as well as active or dormant insects enclosed in the package. Many packaging materials have been evaluated over a period of time in package form but no material has yet been found that is completely resistant to penetration. Many are highly resistant, however, and are only penetrated after subjection to extremely severe conditions. Insects will generally only attack packages that contain foods attractive to them. Their activity is almost wholly regulated by temperature and they are killed if the temperature is reduced to 30 to 40°F for extended periods or if the temperature is raised to 120°F for 1 hour or more. Their optimum growth seems to be in the range 70-95°F and their life cycle is usually about 4 to 6 weeks. It is their ability to multiply that accounts for the serious damage they can inflict once they are inside a package. The problem of keeping them from inside is as much a sanitation problem in the packaging operation, with a thorough sterilization of the contents, as it is a protection of the package from outside infestation.

Many tests have been run to determine what compounds are most effective in repelling insect attacks; dinitrophenol derivatives and DDT have shown the most promise. At least 5% DDT is recommended for use in coating, and care must be exercised when using wax or hot-melt coatings because its effectiveness is diminished when it is held at high temperatures. The best over-all recommendations for this type of protective packaging are as follows:

1. See that all materials to be packaged are fumigated or otherwise sterilized before packaging.
2. Use a streamlined package as free of folds and creases as possible and completely sealed in every respect.
3. Incorporate DDT to the extent of 5% in the coating whenever possible.

Almost any package properly made and sealed will protect the contents for a long period of time if both contents and package are sterilized and treated.

Low temperature (below 65°F) and low humidity are important in preventing external attack of the package. High humidity and 70-95°F temperatures present optimum conditions for insect penetration.

Many physical conditions of the package also influence penetration. A smooth overall surface of the package with no crevices and the minimum of seam, makes it difficult for an insect to obtain footing and a point of attack. The thicker the packaging medium and particularly the coating, the more difficult it is to infest, since small insects cannot bore into a thick coating and it is correspondingly more difficult for the larger ones to penetrate. Surface hardness is important for obvious reasons. An abrasive layer makes insect penetration extremely difficult. A wax double dip has been very successful as an insect barrier.

Rodent Resistance

The question of rodent resistance is an unsolved problem at the present time. No single flexible barrier effective against rodents and no chemical is available commercially that is satisfactory as a rodent repellent or rodenticide. Some proprietary rodenticides, such as 1080 and Antu (α-naphthyl thiourea), are quite powerful, but generally they cannot be used in packaging materials because they are also toxic to man and domestic animals.

Technology of Fungicidal Wrapping Materials

Packaging materials may consist of foil, films, or plain or treated papers. Foil will not support the growth of molds unless the surface is contaminated in some manner. Films vary in their resistance to fungus growth, and the cellulose acetate, cellulose acetate butyrate, ethyl cellulose, polyethylene, Saran, polyvinyl chloride and polyvinyl chloride-acetate copolymers, and Pliofilm all appear to be fungus resistant. Polyvinyl acetate shows some resistance, but cellulose nitrate supports moderate growth. In many cases such growths may be due to the plasticizers if they are not resistant in themselves. Scrim cloth does not offer any mold resistance, nor does MSYT and PT type of Cellophane. MST and MSAT both exhibit good resistance.

As a paper base, sulfite and kraft are both used; kraft seems to possess more inherent resistance than does the sulfite. Glassine and parchment both support growth, although the parchment exhibits more resistance. It is possible to sterilize the packaging paper before use, and sterile conditions should exist in the packaging plant. Because the air is full of spores or bacteria that are constantly being deposited, even sterile packages quickly become contaminated under normal storage and use. The safest procedure is to incorporate a fungicidal agent into the barrier material.

Many agents have been used, but care must be exercised so that the compound will not present toxicological problems to the user and handler of the package. Calcium propionate is a satisfactory fungicide in that it is odorless, tasteless and nontoxic. Sodium propionate and propionic acid are not toxic, and are efficient but they are not permanent and are limited in their use to materials with a short shelf life. Biphenyl and o-phenyl phenol are good fungicides and also prevent growth within the package; o-hydroxybiphenyl and chlorobenzoic acid are also suitable for the same reasons.

Benzaldehyde, 2-chloropyridine, ethyl mercuric chloride, and o-chlorophenol all exhibit interesting properties, but have not been cleared completely for use in food wrappings. Pentachlorophenol is only slightly volatile, but is only used to protect the outer wrapper as it is toxic. Phenyl mercuric compounds are excellent fungicides but are also toxic and cause skin irritation. Salicylanilide is an effective fungicide—nonvolatile, colorless, odorless, and reported to be nontoxic. A special moldproofing agent is made by adsorbing silver on collodial carbon for incorporation into plastic coatings. The coatings are dark in color, but are nontoxic, odorless, and tasteless.

Some fungicides perform well in some coatings and yet fail in others. Chlorinated phenols are among the best agents used in casein coatings (Figs. 5-2 and 5-3), yet they often fail badly in other than protein-

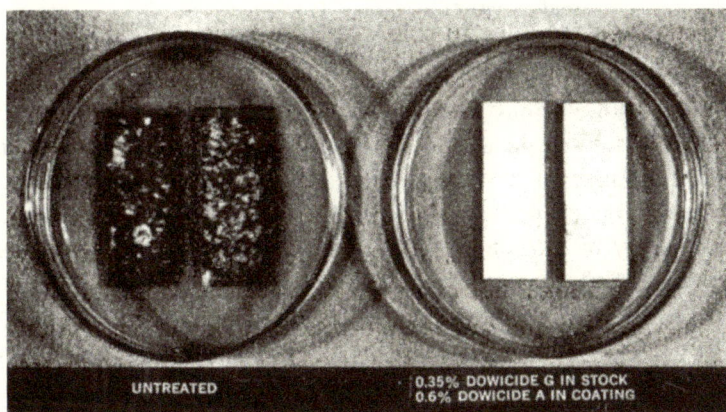

Fig. 5-3. Effect of varying the fungicide concentration of various agents in casein-coated paper on the fungistatic properties of the sheet (*Courtesy of Dow Chemical Co.*).

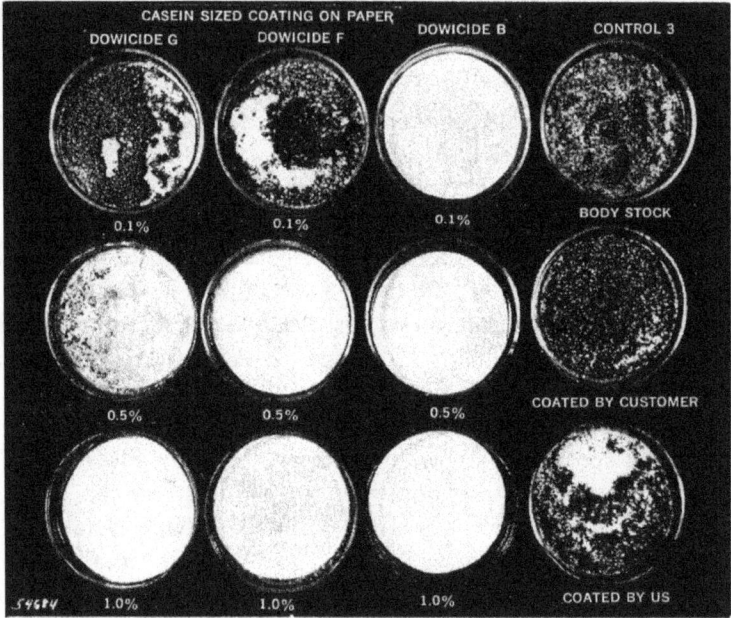

Fig. 5-2. Effect of using fungicides in casein-coated paper for preventing mold growth (*Courtesy of Dow Chemical Co.*).

base coatings. Various mercurial-base compounds have been found to be satisfactory in protective coatings, but can only be used in small amounts because of their toxicity, and thus are mainly limited to outer protection on packages. Mercury oxide, mercury chloride (calomel), mercury bichloride (corrosive sublimate), as well as organic salts of mercury such as mercury naphthenate, pyridyl mercury stearate, phenyl mercury naphthenate, and phenyl mercury oleate, all have shown definite promise. These are effective against fungi, but also have some bactericidal action. Certain investigators claim that long-chain mercurial compounds are safe for use with most foods, even when they contain large amounts of water because their solubility in water is so negligible. These compounds can be coated on the inside of the barrier in concentrations of 1/1000 to 1/5000. They do a satisfactory job as a fungicide and are perfectly safe with respect to toxicity. Solvent-type coatings with pentachlorophenol and salicylanilide produced fair fungus resistance whereas a plain paraphenyl-phenol-formaldehyde varnish offered good resistance to fungi.

In general, it is quite possible to produce functional packaging materials that possess good resistance to fungi and bacterial growths. Foil and certain films are resistant in themselves, and paper combined with suitable coatings, laminants, and saturants also possess the required resistance without any problem of odor, taste, or toxicity. The base paper can be treated in the beater or on the paper machine.

FLAME-RETARDANT PAPERS

The flameproofing of papers, cloth, and films is a problem that has occupied individual researchers and research organizations for many years; considerable literature and many patents cover the subject. As a result of all this work, however, few really effective new agents have been developed.

Three basic materials produced by the converting industry require attention from the standpoint of flameproofing. Foil is not considered since it is a fireproof material and will not burn.

1. The plastic and cellulose-base films offer a wide range of combustible properties. Some, such as nitrocellulose and polyethylene, are inflammable; others, such as treated Cellophane, ethyl cellulose, and cellulose acetate, will burn. The vinyls, Pliofilm, and Saran are nonflammable. A great deal of work has been done on the use of special plasticizers that possess flame retardant properties; most cellulose base films, other than nitrocellulose can be formulated so as to possess a certain degree of flame retardency.

2. The printing or coating material, the laminant, or the saturating agent used in the finished sheet must be made flame resistant or at least flame retardant. Many flameproofing agents, fire retarding plasticizers, and noncombustible synthetic resins are suitable for this purpose. It is beyond the scope of this work to go into the details of the subject.

3. The base paper or cloth, is cellulosic in nature and will burn. It has the further disadvantage that it is in an ideal physical state to burn, since it offers a great deal of surface area for a given weight. The flameproofing of the paper is the converters' problem and an understanding of flameproofing methods is of basic importance.

Fundamentally, no material can burn unless its temperature is raised to the ignition point in the presence of oxygen or oxygen-containing compounds. The material must decompose and liberate combustible gases; these gases are first to be ignited when burning occurs. Chemical instability alone will not cause combustion; it is also necessary that the liberated gases be inflammable. It is apparent that in any consideration of flammability the chemical nature of the materials plays a vital role. Cellulose is inflammable and it is practically impossible to combine another material with it to make the resulting product flameproof or fireproof. It is possible, however, to treat the paper so as to make it difficult to ignite. Such a treated sheet is called "flame retardant."

The mechanism of flame retardency is still a debated subject. Certain flameproofing agents, such as ammonium chloride and borax, have been used for many years, but just how they protect the sheet is not clear. Fordyce describes the following three concepts:

1. The first explanation is that ammonium salts and the like decompose to give off noninflammable gases, such as ammonia, and exclude oxygen from the paper. If this were the entire story, however, certain other salts such as carbonates and bicarbonates would be expected to be much more effective than they are.

2. A second group of investigators postulates that a low-melting material such as borax melts at elevated temperatures and spreads out over the surface as a thin protective skin, preventing free access of oxygen to the combustible paper material. In the case of borax this is probably true.

3. The last possibility is that the flameproofing agent alters the course of the combustion, i.e. during rapid oxidation less inflammable gas is produced and more carbonaceous (charred) material and more water are formed. In the case of acidic materials such as diammonium phosphate and ammonium sulfamate, a reaction is possible with cloth or paper, tending to support this last theory.

All of these factors are important, but it is also possible that there may be cases where one of the three is preponderant to the almost exclusion of the other two, while in other cases the three may possess equal importance. It must be emphasized again that cellulosic materials treated with these compounds are flame-resistant as tested by standard methods and are not fireproof in the sense of foil or asbestos. All will burn if heated to high enough temperatures. The effect of the flameproofing or fire-retarding agent, regardless of the mechanism, is to reduce greatly the rate of combustion of the treated material and to reduce afterglow to a minimum.

Flame-Retarding Agents

Some of the oldest flameproofing agents have been previously mentioned; these are soluble salts such as borax and ammonium chloride. Other soluble flameproofing agents are diammonium phosphate and diammonium ethyl phosphate, both of which are applied from water solution and require about 15% on the weight of the paper to give adequate fire protection and prevent afterglow. Another agent that can be applied from a water solution is a boric acid-sodium borate mixture and this must be added to the sheet in amounts of 25% to obtain fire retardency, but it does not protect against afterglow. A more recent development in this class is ammonium sulfamate, which is added in amounts of 15 to 20% by weight to the sheet and gives excellent flame-retardancy and fair afterglow characteristics. These agents all have the disadvantage of being water-soluble, and any severe wetting reduces or eliminates the retardant characteristics.

A different type of flame retarding compound is formulated from chlorinated paraffin or antimony oxide. This agent is water insoluble, and imparts excellent flame resistance, good waterproofing, and fair afterglow properties. When antimony oxide is used as the sole flame retardant, it is only effective when the amount of organic matter is less than 50%, i.e. when the pigment is 50% of the sheet. The use of chlorinated paraffin is found to be most effective in the presence of antimony oxide. Chlorinated rubber applied from a toluene solution has proved highly successful.

Other agents suggested by various investigators are ammonium sulfate, ammonium carbonate, aluminum hydroxide, zinc sulfate, magnesium sulfate, sodium phosphate, zinc chloride, and urea.

Zinc borate is a new retardant developed during World War II. This compound has good fire-retardant properties and also exhibits water resistance and a definite fungicidal action against common cellulose-destroying fungi.

Nitrogen-phosphorus compounds are the basis of recently developed compounds, which are still in the experimental stage. They are claimed to retard flame propagation by chemical suppression of the flammability. The compounds owe their effectiveness to the expansion of the resin on exposure to heat and the protection against combustion of the charred frothy mass by the ammonium phosphates that are present. Such compositions can be made from urea-formaldehyde resins and ammonium phosphate.

According to Jones, nearly thirty-five chemicals are good fire-proofing agents. These are mainly inorganic in nature, but a few organics are also used such as hexachlorobenzene in o-dichlorobenzene. The compounds used most frequently are chlorides of ammonia, calcium, zinc, copper, and magnesium; sulfates of ammonia, nickel and aluminum; sodium arsenate; boric acid and phosphoric acid. Some such chemicals are used alone and others in mixtures, such as 70 parts of borax mixed with 30 parts of monoammonium phosphate. Mono- and diammonium phosphates are excellent as they retard flame and afterglow even in low concentrations, have no corrosive action on metals, are not hygroscopic under normal working conditions and, if well impregnated, possess a retentivity of 75% after water leaching.

A new type of fire retarding coating, developed at the Forest Products Laboratory, is composed of equal parts by weight of monoammonium phosphate and a 2% solution of sodium alginate. Another recommendation is a mixture containing 50 parts of monoammonium phosphate, 5 parts titanium dioxide, and 45 parts of a 2% solution of sodium alginate. Other coating compounds containing borax, boric acid, phosphates, clay, as well as other suspending agents such as methyl cellulose, casein, and synthetic resins, were also suggested.

The major weakness of most of the flame-retarding agents discussed in this section is their inability to retain their effectiveness after exposure to water, because of the leaching out of the water-soluble components. So far as is known, no water-insoluble compound has been found equal in its effectiveness to such water-soluble compounds as ammonium phosphate, borax, and sodium silicate. Water-insoluble compounds that possess flame retarding characteristics are zinc borate, chlorinated rubber, chlorinated paraffin, antimony oxide, and potassium pyroantimonate.

Plasticized Papers

G. SCHMIDT

The modification of the paper web by natural or synthetic materials of resinous or elastic nature has not always been successful without some greater or lesser change in the bending or folding characteristics of the untreated fiber mat. Most of the papers used for coating or impregnation can be considered stiff, rattly, or having resistance to bending when applied in certain functions. These characteristics can be modified with certain materials to produce limpness, softness, or low resistance to bending. Such alteration of paper is commonly called plasticization, and the materials are designated as paper plasticizers.

The theory explaining the mechanism of cellulose fiber plasticization is most controversial and has no place in this short discussion. For all practical purposes, however, the only successful plasticizer of paper fibers is water. Water is the only lubricant that can cause each fiber in a paper to bend with less resistance than in the dry state. Surrounding the comparatively stiff dry fibers of paper with an elastic or plastic material does not in any great degree cause a change in limpness of the paper such as water does.

Cellulose fibers are somewhat hygroscopic, but the amount of water attracted at normal humidities by them is quite insufficient to make a noticeable change. Other hygroscopic materials having a greater hygroscopicity must be brought in contact with the fibers, so that the water they attract can be transferred directly to the fiber.

Glycerin was the first and practically the only material used to plasticize paper during the years preceding World War II. Some work in the use of invert sugar, either with glycerin or alone, had been progressing, but the evident changes wrought by a small amount of glycerin in comparison to a large amount of sugar made the glycerin first choice. Before the war, the application of these materials to paper could hardly be called important, for only certain highly hydrated, poor-folding grades were treated. Some glassine and parchmentized grades were treated with glycerin, as were the first paper drapes and shades.

Widespread substitution of paper for other materials was necessary to accentuate the importance of the changes that can be obtained through plasticizing. Without the incorporation of the characteristics of folding, bending, or being somewhat limp, many treated papers would fail in their end use. The solution of the problems of plasticizing has not been accomplished without much research. The old stand-by humectant, glycerin, was found to be unsuitable in many cases. Its vapor pressure being somewhat high, glycerin evaporated from the paper too rapidly and its humectant properties at low atmospheric humidities left much to be desired. Certain surface coatings on a glycerin-plasticized sheet were found to absorb the glycerin from the paper, rendering the coating tacky or too plastic. Some coatings reduced cohesion to the glycerin-coated fibers. Invert sugar was found to be less efficient than glycerine and of doubtful use below 30% relative humidity.

Research work in the field of paper plasticization has resulted in a considerably greater range of compounds adaptable to the art, each of which has its own sphere of use. Of particular importance have been the various glycols and polyglycols. These water-soluble or dispersible chemicals have wider humectant ranges than glycerin and have vapor pressures such that they remain in the paper for considerably longer periods. Certain classes of the polyethylene glycols can be formulated to resist migration into resin top coatings or impregnants. Certain hygroscopic salts have also been found to be good paper plasticizers. Potassium acetate in the presence of small percentages of ammonium acetate or other modifiers has been used to a great extent. The added flame retardant properties of the acetates have been helpful.

In recent months, amine-sulfamates have been given a great deal of attention, particularly because of their excellent plasticizing efficiency and their inherent flame-retardant properties. The desirable characteristics of this salt as a low-humidity humectant are of added importance. Many other chemicals have been investigated as possible paper softeners, but because of their cost, nonavailability in quantity, or their effect on the paper fibers, they have been pigeonholed or discarded. Most of the new humectants have been judged toxic and cannot be used in papers in contact with foods. Glycerin and propylene glycol are not in this class. Some of the polyethylene glycols may be classified as nontoxic materials, but they are not accepted by the Food and Drug Act recommendations.

Paper plasticizers are usually applied after the paper web has been formed and dried. Since most of the chemicals are water-soluble and applied from a water solution, wet strength sufficient to hold the fibers together is usually necessary. Size tub, diptank and squeeze roll, spray, or fog are the usual methods of application. The paper may be subsequently dried, or it may be rolled up in a damp condition, depending upon its further treatment. All papers suitable for impregnation with waterborne chemicals can be plasticized. Wetting agents compatible with the plasticizer solution aid impregnation into sized or low water absorption papers, but unsized webs are in the most satisfactory class.

Not all plasticized papers are so heavily treated that the physical change is easily detected. High levels of treatment in ordinary papers usually result in strength reduction or sweating at high humidities, both of which may be detrimental to the end use. In some applications of plasticizing such as with greaseproof papers, the decreased resistance to fold attained with a few pounds of softener per ream can result in high folded grease resistance with lower coating weights. Papers coated with resins, which stiffen the sheet, can be brought back to their original flexibility with a low level of plasticizer treatment. But some applications, such as in paper drapes, require very high levels of treatment and the flame retardant properties of certain of the plasticizers are directly applicable in this case. Noiseless candy bags, barber throws, hospital sheets and pillow cases are further papers that require plasticizers, but only those softeners that are noninjurious to the skin or are nontoxic can be used.

Plasticizing has gained greater importance because of these wide and diverse applications of paper, and with research in the field progressing at its present rate, the ideal paper plasticizer may soon be found. It will have plasticizing action in itself, and will be independent of atmospheric humidity. A plasticized product will then be as soft at $0°$ as at $99°$ relative humidity.

chapter 6

Reprographic Papers

D. S. MOST

HISTORICAL BACKGROUND

In *Specialty Papers*[1] the four most important office duplicating processes were listed as carbon paper, hectograph duplicating, mimeograph, and *Multilith*. Only blueprinting, diazo copying, and the Van Dyke process were discussed in the section devoted to copy methods. Since that date several completely new systems have been created to make the copying and reproduction of existing documents simple, convenient, and relatively cheap.

The need (and also the market) for document copying systems goes back as far as man's ability to record his first messages. One wonders what process would have been suitable for producing copies of cuneiform tablets. Early business transactions had to be accomplished without benefit of multiple copies. In fact, one of the very earliest copying methods reported was based on the use of a high-color-strength ink to prepare the original handwritten document, which was then pressed against the surface of a sheet of dampened tissue paper. Speed, resolution, and convenience must have left much to be desired. Cost per copy, however, was undoubtedly low.

The word *reprography* is of recent creation and is generally used to include those systems and processes for producing either single or multiple copies of a given original by any of several methods. It is of broader applicability than *duplicating* or *copying* and has been quickly accepted into the vocabulary of the industry. In 1963 the first International Congress of Reprography was held in Cologne, Germany.

---

[1] Editor, R.H. Mosher, Chemical Publishing Co., Inc., New York (1950).

Papers dealing with a range of subjects in the fields of nonconventional imaging systems and their applications were read and discussed. Reprography also includes the preparation and use of offset plates for short-run printing purposes. Many of the new copying processes have immediate application to the production of such plates.

The "in-plant" reproduction department which once occupied a small corner in the basement or the supply room has become something of a special concern with the advent of the many newer reprographic systems. No longer is the office print shop the province of the forgotten man. In England an official society called "The Institute of Reprographic Technology" was created to deal with the many subjects of common interest to its members. It also publishes a small quarterly journal on the subject of reprographic systems and attempts to develop a more professional character for its members.

The processes of interest in this chapter are those normally encountered in the office copy segment of the reprographic industry. The following list illustrates the wide diversity of processes employed for such purposes:

1. Diazo
   a) dry type
   b) moist type
2. Diffusion transfer (silver halide)
3. Dye transfer (similar in usage to diffusion transfer)
4. Thermographic
5. Electrophotographic
   a) transfer systems
   b) direct systems

Within each of the above major headings can be found several derivative systems in commercial use and a number of applications under intensive development, but not yet commercial.

Research and development is being carried on actively with systems based on photochromic reactions, photopolymerization, vesicular diazo, as well as the more nearly conventional heat-developable diazos that do not require any secondary chemical processing for latent image development. Thermographic systems that employ heat-induced oil distillation from the original onto the copy sheet are under development as are transfer thermographic systems. Systems based on the latter approach are already in use commercially.

Although the glamour systems of today command the major efforts in R & D programs and in commercial marketing, in maintaining perspective one must realize that the first stencil cutting technique credited to A. B. Dick was developed in 1883 and is still going strong today. The original instrument was recently donated to the Smithsonian Institution and is being housed in its collection of significant graphic arts memorabilia.

PRESENT STATE OF THE ART

At this writing (1968) the market for the major reprographic processes is in a state of flux with the newer electrostatic systems rapidly displacing several of the older systems. Photographic transfer systems based on dye or silver salt diffusion are in the declining phase of their industrial life cycle. Despite process automation and simplification, these systems continue to show a steady sales decline.

Diazo systems, on the other hand, continue to show growth even if not at the same spectacular rates as the new electrophotographic systems. Methods that simplify the development of the diazo latent image have been introduced commercially and these have assisted in placing diazo machines into offices and applications where they were previously not found.

Thermal processes continue to be improved and still hold a relatively significant portion of the over-all copy market. Most authorities, however, expect a decline to occur in the demand for the older thermal copying processes. Improved systems that eliminate the characteristic color blindness of thermographic copiers have been developed and are being marketed successfully. Better types of paper are also used to produce the final copy. Until some as yet unannounced process makes its appearance, the bulk of present generation office copying systems with be based on electrophotography.

As a separate, identifiable market, office copying is expected to reach the $1 billion level by 1970. If one were to include reprography in general, thereby bringing in the value of machines and supplies used in duplicating and in-plant printing, the figures increase substantially. An executive of the Xerox Corporation* reported in 1964 that 0.3% of the duplicating market totaled more than the 9.5 billion copies produced by all copiers and copying processes that year. We do indeed live in an age of mushrooming paper work and red tape. Our annual consumption of paper and paper products was close to 500 lb *per capita*

* C. P. McColough, President, Xerox Corp., at a talk given to the Los Angeles Society of Financial Analysts, June 3, 1965.

in 1965, much of which was consumed in reprographic applications.

Some data are of interest in considering the full magnitude of the situation. The Hoover Commission in studying government procedures made an estimate of "over 1 *trillion* (10^{12}) pieces of paper filed and stored annually." In addition they estimated that 1.75 billion pieces are being added to files annually. Stanford Research Institute in a study entitled *The Office Paper Explosion* reported that the 3,000,000 volumes currently stored in Yale's Sterling Memorial Library are increasing at the rate of 100,000 per year. Furthermore, of the 20 million documents that Chemical Abstracts has abstracted since 1907, fully 40% were generated during the last 10 years. Stanford estimated that the world output of such scientific information is growing at the rate of 5% per year, therefore in just 15 years the total output of such information will again double.

The full implication of these numbers is both frightening and fascinating. Means must be developed to cope with and digest such staggering amounts of information. Record keeping and document storage and retrieval systems will be developed to cope with this flood of knowledge. To this end, electronic data processing is rapidly being applied to paper handling and storage problems and will rapidly become the heart of the best information storage and retrieval systems. For example, IBM has recently announced a microfilm storage and retrieval system capable of storing up to 504,000 microfilm images and delivering any one on demand in six seconds. The output of the machine is a duplicate microfilm that can either be read on location or reproduced as an enlarged hard copy. If greater capacity is required, the system is available with special shelf storage units capable of handling an additional 2.5 million images that can be loaded into the system as required. The machine is also capable of keeping track of these additional images.

Generally speaking the output of such EDP systems will include a hard copy printout capability. Mechanical printers in use today on computer outputs are tending to slow down the over-all speeds of these systems, which are electronically capable of generating data substantially faster than the printers are capable of printing them. Printout systems that employ other forms of energy will be adapted to more nearly meet the requirements of this new generation of equipment.

SPECIFIC REPROGRAPHIC SYSTEMS

Diazo Processes

The oldest, and in many respects the simplest, reprographic process still in active service and showing steady growth is the diazo process. It is an application of dyestuff chemistry based on the strong color-forming coupling reaction between diazonium compounds and substituted phenols in an alkaline medium. The initial work was done in Germany in the early 1920's with the first commercial product marketed by Kalle, A. G. in 1924.

In its simplest manifestation a typical "dry" developing diazo paper is coated with a complex mixture of chemicals designed to produce the greatest sensitivity to ultraviolet light consistent with the grade and intended application of the paper. Ultraviolet light is used to decompose the diazonium compound in the nonimage areas of the copy paper. As the coating ingredients are laid down from aqueous solutions onto the web surface, the control of pH is essential to prevent premature coupling between the color-forming constituents.

Fig. 6-1 illustrates the basic steps involved in the production of a diazo image.

The chemical reactions involved are illustrated in Fig. 6-2.

In terms of photographic speed the decomposition reaction proceeds relatively slowly and hence requires fairly powerful UV light sources. Also, only one-sided originals printed on translucent substrates can be copied. The pH change necessary for dye formation in the so-called "dry" diazo process is brought about by ammonia fumes and by contact with alkaline developer solutions in the so-called "moist" process. The latter process also has the potential advantage of simplifying the coating required on the paper because the coupling agent can be included in the alkaline developer solution.

In recent years completely dry developing diazo papers that contain the developing agent as a component of the coating formulation itself have been introduced. The exposed sheet becoming the latent image is passed over heated rollers to liberate a base and cause the coupling reaction to take place. However, the limitation of one-sided originals still holds.

Coating Technology

Base stocks for diazo coatings require combinations of special

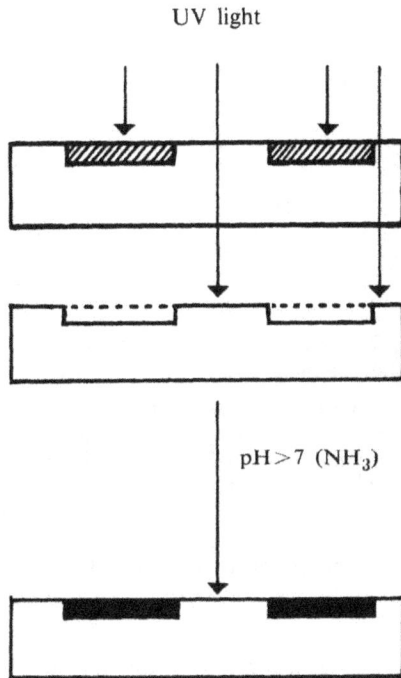

Fig. 6-1. Production of diazo image. (a) Original on translucent paper; (b) diazo paper, coating pH—3; (c) developed image. At (a) the copy paper is exposed to a strong UV source shining through the original. Where the UV passes through the original and strikes the coating, the diazonium compound is decomposed and rendered inert. In (b) the pH of the coating is raised, causing the color-forming coupling reaction to occur very quickly and producing in (c) a positive copy of the original.

properties. Iron content of the paper must be low and should not exceed 20 ppm. In addition, the formation and caliper must be extremely uniform because of the very low weight of sensitizing formula applied. Sizing must also be carefully controlled to eliminate significant variations in the absorption properties of the stock. Freedom from curl is also essential.

Because diazo coatings are applied from solution, great care must be taken during the production of the formulations themselves. Tile tanks and stainless steel mixing equipment with good filtration and generally excellent cleanliness are required to obtain optimum results from a diazo formulation. Inasmuch as the total salt content

$$N\equiv N^{\oplus}$$

(a)

**diazonium salt,
colorless**

(b)

**diazonium   coupler,   dye image
compound phloroglucinol**

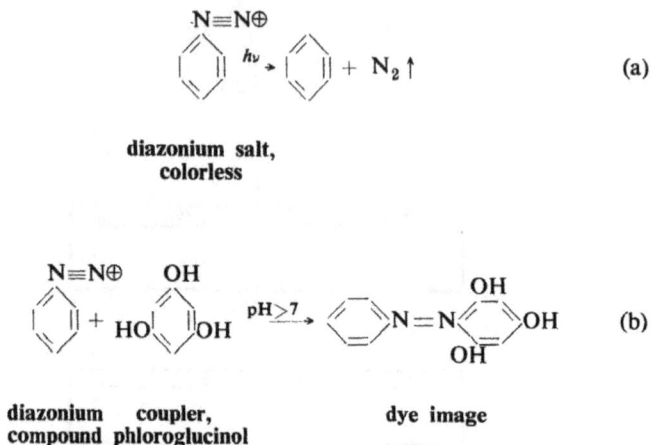

Fig. 6-2. Chemistry of diazo image formation. In (a) the UV light decomposes the diazonium compound in the coating, leaving a latent image of unreacted diazonium and coupler. In (b) the pH is raised to over 7 and coupling occurs with the formation of a bright dye stuff.

of the solution water is quite important, only distilled or demineralized water is used. For most purposes demineralized water is adequate.

The pH of dry developing diazo layers is held at quite low values (2-3) and is buffered to insure stability. When such layers are contacted with ammonia fumes, the pH is quickly raised to the alkaline side and the dye-producing coupling reaction occurs.

For papers to be used in solution-developing machines such levels of acidity in the coating formula are not necessary to maintain. This is because the coupling compound is only brought into contact with the coating when the sheet is being developed to produce the final image. The alkaline developing solution contains the coupler as well.

In the case of heat-developable diazo papers the diazonium compound, the coupler, and a base releasing material (e.g. urea, ethyl urea, or guanidine) are applied either as stable coatings activatable by heat or in separate layers to prevent premature coupling. These solutions must be prepared free of any active alkali. The patent literature contains many examples of these base release systems.

Precoatings of diazo base stocks, a relatively recent improvement in the art, has led to better density images, greater contrast, sharper details, and generally improved image quality. Precoating procedures involve the application to the web surface of aqueous dispersions of colloidal silica or other high surface area pigments with or without binders. Such binder/pigment coatings may include clays, talcs, or silicates in addition to or in place of colloidal silica. When applied to the coating side of a diazo base stock, these materials present a new type of surface as far as the diazo reactants are concerned. The fine pigment particles adsorb the formula components and concentrate them on the very topmost regions of the paper surface. Therefore, when the image is developed maximum color brilliance and color density result.

Modern blueprinters also benefit from the use of precoated stocks and the difference between prints made on precoated as against non-precoated stocks is great. The former prints show brighter, more brilliant blues relative to the latter which tend to be on the dull and rather flat side.

Precoats are applied to the web by roll applicators, which are followed by metering rods or scraper bars to doctor off the excess. Air knives (and trailing blade or gravure coaters) can be used to apply such precoats. The sizing of the base stock must be carefully controlled to avoid having the precoat formulation strike into the web and leave the surface deficient in pigmented coating.

Because diazo coatings are laid down from aqueous systems, the problems of unequal sheet stresses during drying are real and must be dealt with constantly. Modern diazo coating technology employs multiple head coaters that can apply the precoat, the sensitizing coat, and a curl-compensating back coating in one operation. Drying is accomplished by use of high-velocity air jets and subdued lighting is used in all areas wherein the coating may be exposed. Moisture content of the finished paper is quite important and is generally in the range of 3-5% depending on the specific application. When high moisture is left in the web storage, stability is adversely affected. If the paper is over-dryed, subsequent developing by ammonia vapors is materially retarded. Therefore good instrumentation for determining moisture contents of these grades is essential.

Production speeds of diazo coaters are in the range 150-600 ft/min.

A typical diazo formulation contains chemicals to fulfill the following basic functions:

A diazo compound, e.g. p-diazo-diphenylamine sulfate, a coupler, e.g. 2,3-dihydroxynapththalene 6-sulfonic acid sodium salt for color formation, citric acid for pH control, saponin for wetting and spreadability, thiourea as an antioxidant, solubilizers to assist in dissolving the diazo compound, glycerol or other humectants, and alum. Individual formulations, of course, differ markedly in their specific composition but basically each formulation must be built to achieve a specific end result and have acceptable shelf life and storage stability for the wide range of climatic conditions likely to be encountered in the field.

Economic Significance

The 1965 value of diazo machine and supplies sales is estimated at $110 million. In 1964 some 85,000 diazo-type machines were probably in use. The production of coated papers, films, and foils for this purpose was some 1.01×10^9 square yards in 1964 with a value estimated at about $86 million. Therefore, although diazo systems are not among the fastest growing of the reprographic techniques, they do represent a substantial market with a heavy capital investment in machines.

As a market for paper suppliers, diazo and blueprint coaters required about 111,000 tons of chemical wood pulp papers in 1963. In addition some 12,000 tons of rag paper also were employed.

Machine sales alone in 1965 were estimated at $20 million and are projected to some $70 million in 1970 by Stanford Research Institute. By comparison, the figure for 1960 was $10 million.*

In terms of unit copy costs, diazo systems rank with the lowest. Copies on an $8\frac{1}{2} \times 11$ in. basis cost approximately one cent or less for supplies. At the present time the bulk of diazo usage is in the engineering drawing reproduction field and only a small percentage is involved in office copying work. The extension of diazo copying to systems applications, however, is causing changes in this area also. In addition to this positive factor the introduction and continued success of GP® (General Purpose)† bond for controlled master production is helping to expand the market for diazo copying. This type of paper is sufficiently transparent to permit using it for all master preparation for diazo copying machines. It is sufficiently "bond"-like in its over-all characteristics to be useful also as letterhead paper.

Because of the relatively high costs of multiple electrostatic copying some producers have attempted to sell diazo machines to be used

---

* *The Office Paper Explosion*, Stanford Research Institute, Menlo Park, California (1964).

† G. P.® Registered trademark, National Association of Blueprint and Diazotype Coaters.

alongside electrostatic copiers for multiple copying purposes. The first copy is produced electrostatically and subsequent copies are run off on the diazo machine thus using the electrostatic copy as the new master (providing the proper translucency copy paper is employed.) Market experience with this dualistic approach has not been extremely successful.

In general, diazo copying technology continues to show health in the face of newer, more glamorous copying systems, and should continue to enjoy at least a growth rate parallel to that of the national economy. Should any breakthroughs occur to increase materially the light sensitivity of these materials, they would then be in a position to penetrate markets and end uses not now open to them.

Figure 6-3 illustrates the general configuration of a smaller type diazo process machine for routine office use.

Fig. 6-3. Bruning 120 diazo copying machine (*Courtesy of Charles Bruning Co.*).

Silver Diffusion Transfer

Background

The first practical and simple office copying system introduced in this country was based on silver diffusion transfer. The development originated with the Agfa Corporation in Germany and represented a substantial breakthrough in the art relative to the cumbersome, slow, and expensive Photostat® methods, which produced negative silver prints of the original. In the early 1950's a number of American producers introduced machines based on this system and these met with almost instant market acceptance. For the first time even the smallest offices could afford to own such machines and produce good copies at costs around 10 cents each.

Technology

Diffusion transfer systems generally embody a number of known photographic mechanisms. Two sheets of special paper are needed to produce the final positive copy. The negative sheet, which is exposed to light while in contact with the original, is coated with a special gelatin emulsion of a silver halide. When light strikes the non-image areas of the original it is reflected back to the negative sheet and produces silver nuclei on the silver halide crystals. These are the typical silver metal nuclei that act to promote the further reduction of the silver halide crystals to metallic silver and result in the typical high amplification factors known in the art. Those regions of the negative sheet that were adjacent to the information on the original do not get sufficient reflected light to form such nuclei.

To produce the final image, the sheet bearing the negative latent image is placed in face-to-face contact with the positive or receiving sheet. This sheet is not emulsion coated but is coated with a small amount of a so-called nucleating material, which can be metallic silver or a salt that forms a difficultly soluble silver compound (e.g. zinc sulfide). The two papers are then introduced into a developer tank in such a fashion that their surfaces are wetted thoroughly with the developing solution. The developing solution is a mixture of a hydroquinone developing agent and sodium thiosulfate. As the two sheets pass through the developing bath they are brought together in intimate

* Registered trademark, Itek Business Products, Division of Itek Corporation.

(a) *Exposure:*

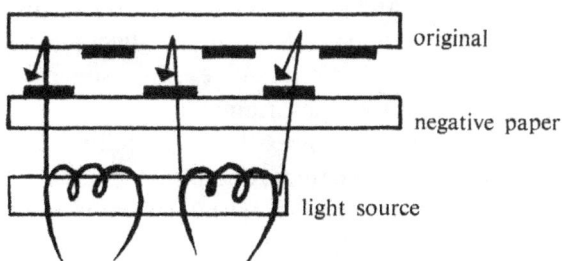

(a) *Exposure:* $Ag^+Cl^- \xrightarrow{h\nu} Ag^\circ(nuclei)/Ag^+Cl^-$ crystals exposed areas of negative.

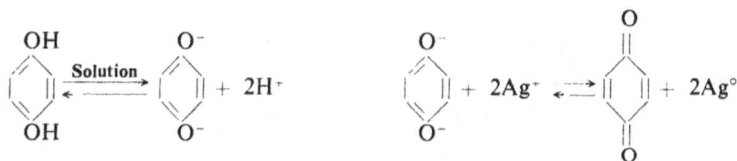

**hydroquinone   active
developer   reducing
                 form**

(b) *Development:* Reduction of silver ions by developer in nonexposed areas of negative.

$$2AgCl + 2Na_2S_2O_3 \longrightarrow Ag_2S_2O_3 \cdot Na_2S_2O_3 + 2NaCl$$

(c) *Solvation of unexposed AgCl:* The ion $[Ag_2(S_2O_3)_2]^=$ is very soluble in the developing solution.

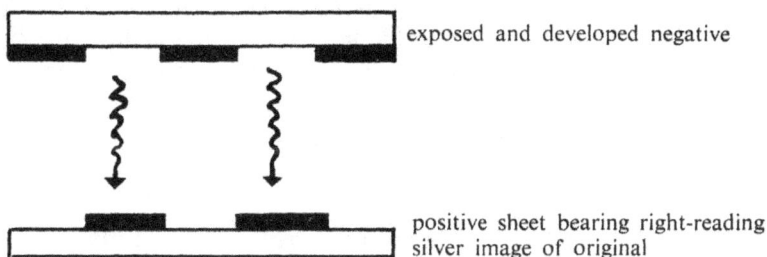

exposed and developed negative

positive sheet bearing right-reading silver image of original

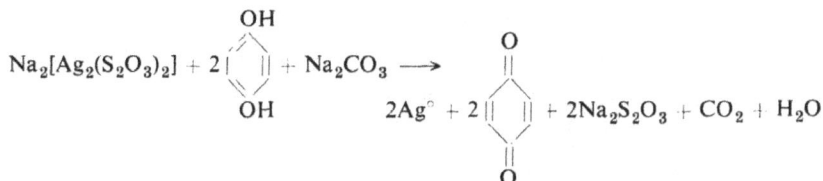

(d) *Diffusion transfer of solvated silver ions and reduction to Ag° in image areas on positive paper.*

Fig. 6-4. Essentials of processes of the Agfa type.

face-to-face contact and are squeezed together by squeegee rolls. When they emerge from the copy machine the two sheets are peeled apart and the negative sheet is discarded, leaving a somewhat damp but good quality positive print of the original.

Schematically, the processes that take place are shown in Fig. 6-4.

The diffusing silver complex ions are prevented from migrating indiscriminately to nonimage areas by the close, pressure contact maintained between the negative and positive papers in the nip of the squeegee rolls. The developing centers on the positive paper materially assist the reduction of sufficient silver ions to produce a dense black image. Equation (d) of Fig. 6-4 illustrates how the thiosulfate solubilizer is regenerated in the process to carry additional silver from the image areas of the negative.

Obviously, with a complex solution such as this the efficiency of the process decreases as more copies are run through the same developer bath, hence it is necessary to have an adequate inventory of de veloping solutions and to add fresh developer for the machine on a fairly frequent basis. Early versions of this process necessitated the daily preparation of fresh solutions because atmospheric oxidation accelerated the decomposition of the developer solution left in the tray and the quality of the copy tended to drop sharply.

The design of these machines is relatively simple, requiring a good light source in either a flat-bed configuration for book copying or a cylindrical or scanning type source for those designs where the original passes through the machine. A timer to control the light exposure is also necessary as is a developing tank through which the papers are passed. Such relatively simple requirements made it possible to de velop low price machines that retailed for less than $100.

Dye Transfer

In response to the market created by the introduction of the silver diffusion transfer process, the Eastman Kodak Company developed and marketed a superficially similar process called Verifax®. This is also a two-step, wet method but it is capable of producing multiple copies from a single exposure of a master or matrix paper.

Verifax negative sheets are coated with a silver/gelatin emulsion containing a tanning developer that causes the gelatin to harden where it is struck by light. The positive or receiving sheet can be ordinary

® Registered trademark, Eastman Kodak Company.

wet-strength paper, but it is generally treated with a surface application of a material such as thiourea to help form a darker final image.

Mechanism of Operation

The original to be copied is placed in contact with the face of the
emulsion-coated negative paper. Exposure is done by the usual reflex
techniques. After the exposure step the negative is immersed in an
alkaline solution containing a gelatin softener. In the light-struck areas
the silver halide is converted to metallic silver and the tanning developer causes the gelatin binder to harden relative to the gelatin in the image
areas that were not light struck. Therefore, the unexposed gelatin
remains swollen and soft. A dye that also develops color is present
in the emulsion. Only the dye in the swollen and softened image areas
can be transferred.

The moist negative sheet is then placed in intimate contact with the
receiving sheet and is passed through squeeze rolls. A thin layer of
the dye-containing unhardened gelatin is transferred to the positive
paper. This transferred layer also contains silver, silver halide, and
developer. The image initially produced by the colored dye gradually
fades and is replaced by a permanent silver image resulting from the
continuing development of the silver halide in the transferred layer.
The negative sheet can be reimmersed in the "developing" solution and
additional copies can be made. They tend to decrease in image density, but recent versions of this process are reported to be capable of
producing up to ten copies. Since the major cost lies in the negative
paper, multiple copies can be produced at low unit cost.

The Polaroid® Land process for producing finished photographic
prints in less than a minute is also a variation of silver diffusion transfer. Both the negative and positive papers are wound on separate
spools within the same film pack and the developer is contained in a
crushable pod located between each film exposure area. As all the
chemicals are left on the print after development, a final acidic wipe is
generally used to achieve permanence and maximum image.

Paper Requirements

The coatings used on the negative papers for both the above processes are special silver halide emulsions, consequently production of
these papers is done by the very specialized photographic coating

® Registered trademark, Polaroid Corporation.

companies. The base stocks are lower cost variations of photographic papers with essentially the same requirements for high purity and good wet strength. Coating is done in very low-level safe light illumination even though the speed of these emulsions is relatively slow. The patent literature is the major source of information on specific formulations and techniques and should be consulted for further details. The bibliography at the end of this chapter gives several good patent references.

The receiving sheet used in diffusion transfer processes has a simpler coating on one side to receive the solubilized silver that forms the image. This coating contains so-called nucleating agents upon which the formation of dark, insoluble silver compounds takes place. Metal sulfides can be used for this purpose. The papers must be quite nonabsorbent as fast drying of the finished print is important. Air drying is necessary after the copy is first produced. More recent papers emerge relatively dry from the developing baths.

Development continues in the area of improving coating treatments for receiving sheets. U.S. Patent 3,203,796 (Gevaert Photo Products Co., 8/31/65) discloses the use of starch ethers as sizing or precoating treatments for papers, onto which are applied the nucleating agent containing coating, which contacts the emulsion side of the developed negative. The claims are that this starch/ether treatment facilitates the removal of the emulsion layer as a coherent membrane, thus improving the quality of the finished copy.

A nongelatin receiving sheet coating is described in U.S. Patent 3,211,551 (Lumoprint Zindler, Oct. 12, 1965) for use with a dry-type diffusion transfer process. In this approach the coating on the receiving sheet is composed of a polyvinyl alcohol and a thermally activatable hydrate (e.g. sodium acetate or citrate) with a polyfunctional alcohol to assure a high residual moisture content and to act as a reducing compound. The receiving sheet is heated and then placed in contact with the exposed silver halide layer, which also contains a developing agent. The heat liberates the water of hydration of the hydrated compound and the silver halide from the unexposed portion of the negative diffuses into the receiving sheet where reduction of the silver halide takes place with resultant image formation. This approach eliminates the need for wet immersion of the papers.

Even the use of prime coatings as simple as starch/clay mixtures are covered in patents for obtaining certain very specific results. British Patent 1,010,202 describes the use of such a subbing layer applied to the receiving sheet underneath the nuclei-containing layer to reduce post-curl of the finished copy after it leaves the developing tray.

Netherlands Application 6,500,536 (A.B.Dick Co., July 19, 1965) discloses coating formulations for the receiving sheets prepared with colloidal silica. An example of such a formulation contains

Polyvinyl alcohol or starch*	—
Colloidal silica (30% solids)	140.0 g
Sodium silicate	7.0 g
Distilled water	70.0 g
Cadmium acetate	0.1 g
Sodium sulfide	0.7 g

* Amount not specified in patent abstract.

This formula is applied to a baryta coated base stock in amounts between 0.1-10 g/m and air-dried. This treatment is claimed to be more convenient and cheaper than gelatin coatings and also makes for easier separation between the print and the negative. In addition, a reduction in the consumption per print of the diffusing silver complex is claimed, thus increasing the number of copies possible from a single negative.

Mead Corp., in Netherlands Application 6,500,727 (July 21, 1965) discloses coating compositions suitable for application by conventional coating machinery. The paper is first coated on both sides with a pigmented composition containing

Clay	50.0 parts
$CaCO_3$	10.0
TiO_2	10.0
Ca stearate	1.0
Na polyphosphate	1.5
Dicyandiamide	4.5

To the above mixture, while it is being kneaded, is added:

$CuSO_4 \cdot 5H_2O$	0.04 parts
Enzyme converted starch	23.00
Styrene-butadiene latex	7.00
$Na_2S \cdot 9H_2O$ (at pH 8.5)	0.015

Solids of the above composition are adjusted to 56% and applied to both sides of the sheet at a weight of 10.5 g/m² (approximately 17 lb/3000 ft²). The paper is given a final top coating (receiving surface) with a formula specified as follows:

Sodium carbonate	1.6 g
Calcium carbonate	10.0
Polyvinyl alcohol	25.0
Ethoxylated starch	25.0

This mixture is heated to 88°C at a solids content of 14% for 40 minutes and then 10 g of an optical bleach are added to the solution at 38°C followed by the addition of an aqueous solution of 0.1 g $AgNO_3$ and 0.02 g $Na_2S \cdot 9H_2O$.

The calcium carbonate containing coating is said to increase the blackness of the image tone of the reduced silver.

Many other examples of special coatings for these papers can be found in the patent literature and these examples are intended only to illustrate the diversity of approaches taken to achieve a desired result.

Economic Significance

In 1963 the combined dollar volume of sales for both these transfer processes was estimated at $120 million. For 1965 the A. D. Little Company estimated a volume of about $95 million. The peak for these processes occurred in 1962 when approximately 1.6 billion copies were produced (A. D. Little estimate). The slope of the consumption curve for these methods turned negative shortly thereafter and is expected to continue declining.

By way of contrast, Electrofax® sales in 1963 were estimated at $30 million, whereas the comparable figure for 1965 was $90 million.

Copy costs have been steadily reduced for diffusion transfer systems with part of this reduction attributable to the fact that the positive sheets can be produced by paper mills on existing equipment. This has tended to eliminate the need for conversion coating of these papers with gelatin. Machine designs have also been improved and simplified to minimize the amount of handling needed with the special developer solutions. Many machines automatically return unused developing solution to storage containers, thus reducing its rate of deterioration.

THERMOGRAPHY

Introduction

In response to the market need for a simpler dry, chemical-free

® Registered trademark, R.C.A. Corporation.

copying system the 3M Company introduced its Thermo-Fax® copying systems in the early 1950's. This was the first reprographic system that required only a piece of special paper and a machine to produce a dry copy. Considering all the handling problems attendant to the use of the then existing processes, the Thermo-Fax approach was revolutionary. The total imaging and developing system was contained in the one sheet of coated paper required for the machine, which in itself was relatively simple and inexpensive.

The concept behind the development of these thermographic copying systems was superbly simple, but the evolution into a commercially acceptable process was quite difficult and costly. If a controlled differential heating between image and nonimage areas on or within a special paper can be produced, then it is only necessary to select or design a chemical reaction or physical change that responds to the available heat to produce a visible color change. Unlike diazo processes or silver transfer processes, such a concept is not tied to any specific chemical system or reaction.

An interesting historical account of the development of the 3M system can be found in *Chemical and Engineering News*, July 13, 1964. In 1940 Dr. Carl S. Miller of the 3M Company first conceived the idea that was later developed into the Thermo-Fax process. He reasoned that if a piece of copy paper containing a thermally reactive system were placed in close contact with a heat-absorbing original then infrared radiation passing through the copy paper would cause the image areas of the original to become heated, the heat would in turn be transferred to the copy sheet by conduction, whereupon a color-forming reaction might be made to occur image-wise in the copy sheet. By using such a direct method he could avoid any reversal of the image and thus produce right-reading images without benefit of any optical systems. Ten years and several millions of dollars later the first commercial system was delivered to the first customer in Washington, D.C., the Central Intelligence Agency. The controlling method patent, U.S. Patent 2,740,896, issued in April 1956, presaged a flood of patents in the field that covered special coatings, papers, machines, and applications.

With the introduction of desk-top sized machines based on thermography, the 3M Company soon sold more Thermo-Fax type machines than were sold for any other office copy process. Other com-

---

® Registered trademark, 3M Co., Minneapolis, Minn.

panies introduced competing equipment sometime after the market had been developed by 3M and a variety of competing papers for use in 3M machines were introduced by independent producers.

Eastman Kodak also has a thermographic system on the market called Ektafax®. This system uses an intermediate for producing a master that is then run through a separate unit with a piece of receiving paper and multiple copies (10 or more) are made from the master. Thermal imaging masters for use with spirit duplicating machines are also now available and a combination machine is marketed today. This machine thermally images an original to produce a spirit duplicating master that is then used directly on the same equipment to produce multiple copies of the original quickly and at low cost.

Fig. 6-5 shows a Thermo-Fax *Secretary* copying machine.

Fig. 6-5. Thermo-Fax Secretary (*Courtesy of 3M Company*).

® Registered trademark, Eastman Kodak Co., Rochester, New York.

Mechanism of Processes

Because thermographic copying is not based on any single or unique chemical reaction, many useful systems that function effectively in the commercial machines available have been developed.

Among the earliest papers that were devised are the so-called "blush" coated sheets. A colored substrate, which is infrared (IR) transmissive is coated on one side with a composition that in its crystallized form is opaque to visible light. When this paper is placed in contact with an IR absorbing original and exposed in a thermal copying machine the coating is melted in the image areas and becomes nonopaque. This allows the dark color of the substrate to show through and produces a usable copy of the original.

A variety of materials has found application in this type of paper. Pressure sensitivity tends to be a problem with such coatings. Examples of these types of systems are:

U.S. Patent 2,710,263 (June 7, 1955)
A black coated underlayer of paper is overcoated with a cadmium stearate dispersion in heptane. This in turn is overcoated with a thin layer of cellulose acetate for protection. A variety of substrates is listed, e.g. glassine, vinyl films, and cellophane.

U.S. Patent 2,859,351 (Nov. 4, 1958)
A non-transparent heat-transparentizable coating comprising cellulose nitrate in a polyethylene glycol.

U.S. Patent 3,125,458 (March 17, 1964)
A coating composition comprising a dispersion of discrete nontransparent particles of a polyvalent metal soap of a fatty acid having at least six carbon atoms in a vehicle, e.g. calcium stearate in ethyl cellulose.

French Patent 1,312,453/81,200 (July 1, 1963)
Describes heat-sensitive coatings based on the application of a layer of cellulose acetate in a mixed solvent system of acetone and toluene containing a heat-sensitive plasticizer such as triphenyl phosphate or cyclohexyl phthalate on a black, coated substrate.

The general technology of hot stylus recording systems formed a reservoir of potentially applicable ideas to be used in thermal copying machines and was drawn upon extensively. The bibliography at the end of this chapter lists a variety of sources for the interested student of such systems.

The so-called chemical papers rapidly gained ascendency in the market because they did not suffer from the serious drawback of pressure sensitivity and unattractive appearance. The patent literature abounds with examples of reactant systems for use in thermographic copying. Dr. C. S. Miller of the 3M Company is the holder of a series of patents that disclose a variety of color forming reactions claimed as suitable for thermal copying papers. The earliest in this series are U.S. Patents 2,663,654 through 2,663,657. The claims in these patents range from the very specific, e.g. the iron salt of a long chain fatty acid plus a phenol, to the very general, e.g. an electron acceptor which is an ionizable compound of an anion and a polyvalent metal cation plus an electron donor which is an ionizable organic chelating agent. The reactants covered by these patents are handled in the form of dispersions of particles of the reactants in IR transparent, nonfusible binder systems.

Because the coating left in the nonimage areas of the copy paper is still heat sensitive, the copies are not "fixed" in the photographic sense. If an imaged sheet of thermal paper were to be reused additional image areas could be produced. In practice the lack of fixing tends to lead to gradual discoloration of the background of the paper. Sunlight striking some papers tends to discolor them badly. Light-fixable thermal systems have been developed, and again the patent literature is the best source of information. One of the reactants is potentially deactivatable by ultraviolet light, for example. Therefore, after thermal imaging is completed the paper need only be exposed to an appropriate light source to render the image fixed by eliminating the reaction capability of the nonimage areas. An example of this type of system can be found in U.S. Patent 2,680,062, July 1, 1954, which discloses the use of diazo sulfonates with azo coupling compounds in thermal copying. The image is first produced in the normal manner and the unreacted diazosulfonate is then destroyed *in situ* by exposing the paper to UV. A variant of this approach is also usable wherein the paper is exposed image-wise to UV, followed by thermal development of the latent image.

Paper Requirements

In the normal imaging configuration wherein the copy paper is placed face up on the original and the two are then held in close contact while they move under an intense line source of infrared radiation, certain initial demands are placed upon the base stock of the copy paper. It must allow a high percentage of the incident IR radiation to pass through itself and strike the letters of the original. If the letters to be copied are IR absorbing, they will then rise in temperature. This temperature rise must be transmitted back to the reactive or fusible coating with a minimum of heat loss and a minimum of lateral migration. For this reason the initial papers selected for this application were of the glassine or greaseproof type. Their IR transmission is high enough to achieve reasonable efficiency and their caliper and apparent density are correct to obtain adequate thermal conductivity.

The disadvantages of these types of papers lie in their tendency towards brittleness and low tear resistance. They also represent a radical departure from the ordinary types of papers found in offices, a factor that led to considerable development efforts to produce more "bond-like" base stocks for thermographic applications. A variety of basis weights became available to satisfy market needs and special systems papers were also developed to produce multiple copies from a single developed master. In this case the multiple copy paper did not need to have the physical characteristics of the usual thermal sheet because a transfer of image was involved. In other types, part of the reactant system was contained on the receiving sheet, which could then be of the more conventional opaque paper character.

Some attempts were made to produce thermal papers by saturating a suitable paper base stock with an aqueous system of reactants capable of being dried *in situ* and later caused to react in the normal fashion. These systems had high potential for low cost and were intended for production by the paper mill itself. When the reactants were distributed through the sheet no problems of coating or cracking or flaking of the imaged layer were encountered.

Coating Technology

Because of the wide diversity in types of reactant systems it is not possible to generalize too broadly. When one is dealing with systems designed to react within a narrow and critical temperature range, great care is necessary to prepare and apply the coatings under very closely controlled conditions.

Processing of the coating formulations depends on the specific system itself. Ball milling of some reactant systems is employed, where as in other solution types of formulations, simple mixing of ingredients is sufficient. In double-coated papers one reactant may be applied in the molten form and the other may be laid down from a solution or dispersion.

The choice of coating hardware is also quite dependent on the specific formulation being handled. Roll coaters of various configurations are used extensively. Gravure, rod, knife, and even air-knife coaters are also useful. Drying control is of course extremely important. Large volumes of low-temperature air must be employed to remove the solvents while avoiding reactions. With coated systems having pressure sensitivity, subsequent handling of the web must be done with extreme caution. In many instances this procedure necessitates the use of specialized center winding equipment.

Coating weights must also be closely controlled because variations in this important parameter can cause variations in the imaging performance of the finished sheet. The finished paper must be stored under controlled conditions to minimize any stability problems. Even though the reactant systems are carefully developed to have definite threshold reaction temperatures, prolonged exposure of most thermal papers to moderately high temperatures produces discoloration.

Fig. 6-6. Dual Spectrum copier (*Courtesy of 3M Company*).

Economic Significance

The 1965 value of thermographic equipment and supplies is estimated at about $100 million. Of this total the 3M Company is be-
lieved to have some 80%. Approximately 1.5 billion thermal copies
were made in 1965 according to Arthur D. Little Company. Recent
improvements in thermal copying systems have acted to arrest the
beginnings of a downturn in the growth curve of this process. The
3M Company has introduced new systems that overcome some of the
shortcomings of the original Thermo-Fax systems and are capable
of producing good quality copies on "bond-like" paper. Small units
are available for the low volume user and automated multiple copy
systems are being marketed for use in the medium volume copying area.

Fig. 6-6 shows a thermal copier recently introduced by 3M.
It uses two sheets of special paper to produce a single permanent copy.

Fig. 6-7. System A-09, multiple copy machine (*Courtesy of 3M Company*).

It is not color blind and can copy from bound volumes.

Fig. 6-7 shows a multiple copy machine that is also based on thermography and can produce up to 25 copies of a given original with a single dial setting. This is a thermal transfer process and the finished copies are on bond-weight paper.

With machines such as those illustrated the growth curve for thermography should continue upward. Paper requirements will also expand accordingly. Such systems are capable of producing spirit masters as well as offset masters, thus making it possible for the user to select his printing system according to the specific needs of a given job.

ELECTROPHOTOGRAPHY

Introduction

In October of 1938 Chester F. Carlson and Otto Kornei produced the first reported electrostatic-type image. It was the forerunner of an entirely new generation of electrophotographic processes and products. The photoconductive "plate" used in their experiments consisted of a sulfur-coated metal surface that was electrostatically charged by rubbing with a handkerchief. It was then selectively discharged by exposure to light through an image-bearing glass microscope slide. The latent electrostatic image was developed on the sulfur plate by sprinkling dyed lycopodium powder over it. The powder on the nonimage areas was blown off and the resulting image was then transferred to a piece of wax paper that was heated to form the final fixed image.

Carlson received his first patent on this process two years later and thus began the long, expensive process of producing a commercially useful system based on electrophotography. Battelle Memorial Institute began working on the system in 1944. In 1948 Xerox Corporation, then the Haloid Co., acquired a license to Battelle's patent rights to the Carlson process and began development work on its own. In January of 1956 Battelle transferred all of its patent rights to Xerox in return for stock in the company and, as a result, became the largest single stockholder in the Xerox Corporation. In 1950 their first commercial unit based on xerography was marketed under the name Xerox Model D Copier. This equipment was used for making xerographic offset plates. In 1960 the "914" copier was marketed and in 1963 the smaller convenience copier called the "813" was introduced. Other machines and systems followed and the success of the Xerox Corporation has already become a business legend.

The approach followed by Xerox continues to be based on the transfer of a powder-developed electrostatic image from a metallic photoconductor surface to a nonsensitized sheet of paper. Heat is applied to the transferred image to fuse it; the resulting copy is generally an excellent facsimile of the original. The early work at Battelle Institute, however, also included research on papers coated with dispersions of photoconductive pigments in resin binders. Xerox did not elect to develop this approach commercially because its marketing strategy was to develop systems that would produce the final copy on unsensitized, plain paper. The advantage of not having to transfer the image from the coated paper to another surface was potentially a major simplifying approach to machine design but Xerox engineers did not feel it important enough to warrant changing the over-all direction of their systems development program. As a result, all present Xerox equipment utilizes the transfer of the developed image from a photoconductive plate surface to an ordinary sheet of well-finished bond-weight paper.

In the early 1950's RCA carried on a program of research in the area of resin/photoconductor coatings on paper and in 1954 introduced their so-called Electrofax® process. This is the process, based on the use of zinc oxide/resin coatings on paper, that does away with the need for transferring the image to a receptor surface. RCA did not elect to produce commercially any systems of its own at the time and instead it licensed the process to a number of companies. Much interest was generated and as a result some 100 nonexclusive licenses were granted to companies ranging from large hardware and systems designers to small independent paper coaters that saw an opportunity to participate in the creation and development of a whole new market area for a specialty coated paper.

Much work and many millions of dollars have since gone into the development of the present generation of electrostatic copy systems and coated papers, but the issue of basic patent dominance is currently being resolved in the courts. By virtue of the earlier Battelle work in resin/pigment coating systems, Xerox was issued U.S. Patent 3,121,006 in February 1964. This patent discloses the use of photoconductive materials such as zinc oxide, magnesium oxide, cadmium sulfide, zinc

® Registered trademark, RCA.

sulfide, and zinc selenide dispersed in resin binders and coated on various substrates including paper. At the present time patent suits and countersuits exist between several of the major companies in the electrostatic machine and supplies industry. The final resolution may include cross licensing agreements that would in effect permit current producers to continue their present lines of development and would redistribute royalty payments now being paid by producers to RCA and Xerox.

Electrophotography Mechanism

Two recent books dealing with the mechanism of electrophotography have been published by the Focal Press of New York. These are *Xerography and Related Processes* by Dessauer and Clark, and *Electrophotography* by R. M. Schaffert. Both volumes summarize the state of the art from the technical standpoint and should be consulted by those wishing to explore this subject in greater depth.

For purposes of this chapter an outline of the basic mechanisms and a discussion of the more pertinent industrial parameters will suffice.

Both xerography as applied by the Xerox Corporation and "Electrofax" as practiced by RCA licensees depend upon the same basic phenomena. A photoconductive insulator is given a blanket electrostatic charge by means of a corona source operating at a relatively high dc potential (ca. 6-10 kV). The photoconductor layer is supported by a conductive backing member, which in the case of Xerox systems is generally aluminum. For zinc oxide coated papers the base stock is made conductive relative to the oxide/resin layer itself. Because photoconductive selenium is a p-type semiconductor wherein conduction occurs through the migration of positive charge carriers or "holes," it is generally charged with a positive ion source. Zinc oxide on the other hand is an n-type semiconductor wherein electrons are the prime charge carriers; hence it is normally charged with a negative corona source. In both instances this charging step must be done in the dark because the layers have their greatest resistivity then and will retain the applied electrostatic charge for relatively long periods of time.

To form the desired electrostatic image the charged layers are then exposed to the desired image via projected light or reflected light depending on the form of the information to be reproduced. Where light of the proper wavelength strikes the photoconductor the electrostatic charge is dissipated, leaving behind an electrostatic charge pattern in the form of the image. Where negative microfilm is involved the image areas of the original are transparent and hence the charge dissipation occurs in the image areas. Special development techniques are employed in these cases to render the discharged areas of the photoconductor visible.

Development of the latent electrostatic image is accomplished by contacting the charged areas with a developer that may either be a solid or liquid dispersion of fine particles carried by another medium. In the case of dry powder development the pigment particles are combined with a carrier such as glass beads to impart an electrostatic charge to the toner particles of opposite polarity to the latent image charges. The oppositely charged developer particles are electrostatically attracted to the image areas. Application of a fusing technique then fixes the dry toner permanently to the photoconductive surface, producing an excellent reproduction of the original.

Xerox equipment utilizes a drum or cylinder made up of a special vitreous selenium layer coated onto aluminum. The dry toner powder is attracted to the latent electrostatic image on the drum and this powder is then transferred to the final receiving sheet. The transfer of the powder is affected by the application of a voltage of opposite polarity to the toner charge sign while the cylinder is in close contact with the paper. The image that then appears on the paper is right reading and is fused to the sheet by means of a radiant heat source in the copy machine itself. The drum is cleaned of residual toner and repeats the cycle. Recently introduced models of Xerox equipment have larger diameter cylinders that make it possible for three successive images to be in process almost simultaneously. Some forty copies per minute of a given original are produced by the Xerox 2400, which utilizes such a large diameter drum. The technology required to affect successful TESI (transfer of electrostatic images) is complex and has been the subject of intensive research and development programs.

Because the selenium drum is likely to be constantly in use in a given system it must not exhibit any serious light fatigue effects. That is, the charge acceptance must not decrease after many successive charging, imaging, and developing cycles. Special alloying and fabricating techniques are required to satisfy this difficult objective.

In the case of *Electrofax* systems based on zinc oxide/resin coatings, the latent electrostatic image is formed directly on the coated surface and need not be subsequently transferred. It is developable by either the dry toner method described above or by a dispersion of toner particles in a dielectric liquid medium. The art of formulating good liquid toners for this application is somewhat difficult but it should be borne in mind that it is generally possible to make a dilute suspension of a black printing ink in a hydrocarbon solvent and obtain deposition of the pigment particles onto the charged areas of the photoconductive coating. Commercially useful toners require much more development effort. Discussions of the theory and practice of toner formulation can be found in the two books on electrophotography referred to on p. 162. In addition, the patent literature discloses a multiplicity of recipes and approaches that may be used.

Because the dry or liquid toner must be fixed to the coated surface directly in *Electrofax* systems, a more careful matching of the coating and toner system must be accomplished. In the case of liquid toners the dispersant or toner vehicle must slightly soften the resin binder of the coating to permit the toner particles to be adhered more strongly when the slightly "wet" coated paper leaves the copying machine. As the dispersant evaporates and some heat is applied to the resin coated pigment particles of the toner, they adhere more tenaciously to the oxide-coated surface and make in effect a fixed copy that is insensitive to light and will remain perfectly readable as long as the oxide coating itself remains intact.

The special requirements of liquid developing copy machines place extra demands on the base stock upon which the oxide is coated. These will be discussed in more detail.

The mechanism as described above is somewhat oversimplified. Each portion of the over-all imaging cycle represents a complex mechanism unto itself. The serious student of electrophotography will understand this and is urged to consult the available literature in the field to grasp more fully what is involved. Much research in electrophotography is being done in Japan and Russia as well as in the U.S.A., hence these important literature sources must not be overlooked. Liquid toner development in Australia has progressed markedly, and in fact a full-color proofing system based on zinc oxide coated papers developed with successive colored toners has been commercially introduced in this country by a licensee of an Australian organization.

Coating Technology of "Electrofax" Papers

Base Stock Requirements

As in every other similar product the production of a high quality zinc oxide coated paper begins with a high quality base stock. Because this application is essentially photographic in character the same considerations of close control, tight specifications, and freedom from possible migrating contaminants are essential.

Base stocks for *Electrofax* applications must have suitable levels of solvent holdout to permit the application of a smooth, uniform thickness of oxide/resin coating. Generally, the pigment dispersion is applied from a solvent medium which may be toluene, xylene, MEK, or naphtha or combinations of these. If the paper has inadequate levels of holdout, penetration of the vehicle will occur and certain of the formula components will drain into the base stock with a subsequent drop in the performance of the finished paper. In developing functional coatings many laboratories first evaluate the qualities of the coating on a substrate such as aluminum foil. This technique completely eliminates the effect of base paper variations and permits formula aspects alone to be studied. Normally, superior results will be obtained for a given coating when it is applied to aluminum foil as compared to the same coating applied to paper. The opacity of zinc oxide coatings is not commensurate with the coating weight generally employed, hence the final color of the coating can be somewhat affected by the color of the substrate.

In addition to the requirement for solvent holdout on the coating side the opposite side of the web must also have high holdout to aliphatic solvents when the paper is to be used in liquid toner systems. Although such machines are equipped with blower systems and sometimes with auxiliary heaters to increase the rate of evaporation of the dispersant film resident on the surface of the developed image, it is essential to prevent migration of the toner into the base stock itself. If this is not done the copy will retain dispersant and may stain surfaces onto which it is placed. It can also cause file drawers to have an odor of dispersant and in some extreme cases could cause a fire hazard. A properly treated base stock with a holdout coating on the back side will not permit dispersant to penetrate the sheet and will permit the squeegee rolls in the developing station of the copy machine to remove almost all of the applied dispersant. Systems builders have critical holdout specifications to maximize the yield of copies obtainable from a given quantity of toner dispersant. Papers that carry out an excessive amount of dispersant from the developing station materially reduce the number of copies obtainable from a given volume of dispersant and toner.

Solvent holdout is usually built into base stocks along with electrical conductivity. Paper normally has resistivities in the region of 10^{10} to 10^{14} ohm/cm^3 depending on relative humidity. Because copy papers must function at both high and low ambient relative humidities, surface resistivity for both liquid and dry toner papers should be of the order of 10^7 to 10^9 ohm/cm^3 over the RH range 75-20%. In actual practice the desired level is achieved at some median point along the conductivity/relative humidity curve and may generally occur at approximately 40% RH. Variations in RH above and below this point cause large variations in surface resistivity, but in actual practice most treated papers coated with a good functional zinc oxide layer give acceptable imaging performance over the 75-20% RH range despite the fact that the surface resistivity of the base stock will have climbed by 2 or 3 orders of magnitude as the low end is approached and may have dropped by 1 or 2 orders as the high end is approached.

The actual surface charge developed under a given set of corona conditions and the distribution of charge between the coated layer and the base stock are also quite dependent upon the apparent surface resistivity of the base stock. Diamond et al. presented an excellent summation of the factors that effect this type of behavior in a paper published by TAPPI at the 1965 Coating Conference. They point out the effect on such critical parameters as dark decay rate (the apparent loss of surface potential from a charged coating left in the dark), rate of charge acceptance, and apparent surface charge as a function of surface resistivity for varying times of charging. In a paper presented at the same conference, W. F. Uhl pointed out some of the practical considerations involved in selecting the proper types of precoats to produce holdout and lower resistivity in base stocks. He illustrated the harmful effects that can occur when a substrate is treated with a material that can migrate into the oxide coating under conditions of high RH. Rosin sizing of base stocks must be carefully evaluated because these materials may be extracted by the liquid toner systems leading to the poisoning of the toner dispersion in the machine. Under some conditions of oxide coating these sizing materials may possibly migrate into the functional coating, causing a marked drop in their over-all performance.

Short of applying the zinc oxide coating to aluminum laminated paper substrates, base stocks must be coated with barrier layers with solvent holdout and conductivity. Commercial products that materially help the paper coater to develop these properties are available. Vaurio[1] presents the results from a comprehensive study of antistatic compounds and special polymer materials useful for reducing the electrical resistivity of paper.

Many barrier coatings are compounded with some pigment content to increase opacity and create a smoother surface for the oxide coating. Some base papers are supercalendered to obtain maximum smoothness and density. When barrier materials that have cold flow properties are applied, calendering can help to eliminate pin holes and insure better integrity of the precoating. Coat weights tend to be kept as low as possible for economic reasons because the commercial conductive polymers currently available are expensive and also because the oxide coatings themselves are in the range of 18-30 $lb/3000 ft^2$. Thus the base stock must be of sufficient weight to handle this amount of coating without being completely limp. Lightweight zinc oxide coated papers will probably begin to appear in the market and developments will lead to an accelerating trend in this direction.

To produce the proper type of electrophotographic base stocks the papermaker must be equipped with some type of coating or sizing equipment either on or off machine. This equipment may be a simple size press or it may be trailing blade, roll, air-knife, or rod-type coaters. Whatever is used should be capable of applying low weights of special coating formulations economically to one or both sides of a given raw stock. As in all coated papers the raw stock qualities are of major importance in determining the quality of the finished product. Compensation for a poor base stock with any type of subsequent coating operation is virtually impossible. In *Electrofax* technology this generalization must always be borne in mind since failure to give proper weight to its significance will only lead to substantially greater problems later on. Specifications on base stocks must be carefully spelled out and should contain limits on the key parameters of solvent holdout, conductivity, caliper, and smoothness.

Coating Preparation

Because present zinc oxide coatings are based on solvent-soluble resin systems the production of suitable coating formulations must

be carried out with explosionproof equipment under closely controlled conditions. Several types of grinding and dispersing hardware
can be used, depending on the specific nature of the individual formulation. Ball mills, Kady mills, sand mills, and ink mills can be used.
Depending again on the specific formulation requirements, a premix
of the formula containing all the resins, pigments, and additives is
prepared. The mixture is then charged into the specific grinding
equipment. The mix is held in the processor until the coating grind
reaches a Hegman gauge reading of 5-7 N.S. Experience will dictate
the exact grind required for each formulation since differences in viscosity, solids, and leveling index, affect the particular grind gauge reading achieved after a given time interval. The objective is to produce a
smooth, uniform dispersion free of agglomerates that would appear
in the final coated paper as lumps or that might cause streaks and scratches on the coating equipment. It is generally advisable to screen the
finished coating ahead of the coater to remove any flakes or lumps of
dried coating that have a tendency to form on surfaces of equipment
that have been wet with the formulation. Screening during recircula-
tion from the coater is usually carried out.

In some instances, dye sensitizers are added after the proper grind
of oxide/resin has been attained. This operation insures optimum
control over the final photographic speed of the coating. Certain dyestuffs used in these formulations may be sensitive to the heat built up
during the grinding or dispersing operation and may not behave properly in the finished coating. By holding the sensitizer addition until
after the dispersion operation one may achieve maximum control of
the photographic properties of the formulation. It is difficult and
often quite expensive to try to remedy functional deficiencies that may
appear in a finished coating containing all of the formulation components.

Quality control tests on finished coating dispersions include such
properties as solids content, viscosity, grind gauge, weight/gallon,
and functionality. The latter catch-all term is the key to the successful
performance of the finished product. The coating is applied to samples
of the selected base stock and is dried under controlled conditions.
Its electrical parameters are then determined by measurement of the
charge acceptance, dark decay rate, light decay rate, and residual voltage. These electrical results may then be compared with the actual
imaging results obtained on a standardized copy machine that has been
set up under constant test conditions. It is important to understand

that when one tests such papers for imaging performance one is simultaneously testing toners and machines also. Therefore, a toner of known performance characteristics must be used with a known machine operating under controlled voltage conditions in an ambient situation of controlled temperature and relative humidity. This empirical approach is necessary to satisfy the stringent control requirements of these papers. Once coated, an *Electrofax* product can only be down-graded in price and applicability if something has not met standards during the preparation of the formula or the base stock.

At the present time the principal photoconductor of commercial importance is French Process zinc oxide in the particle size range 0.25--0.40 micron. It is available from U.S. producers in large quantities and is being introduced into this country by some overseas producers also. Considering the "arty" character of present *Electrofax* papers one must have complete confidence (backed up by close quality control testing) in one's material suppliers. This includes resins, dyes, solvents, and everything else that may go into a given formulation.

Generalizations about resins are difficult to make because the patent literature describes a variety of generic types claimed to produce suitable formulations. Among these are silicones, polyvinyl acetates, alkyds of several types, acrylics, and polyesters. Resin blending is resorted to by many formulators to obtain the complete spectrum of requisite properties and performance in a given coating. As the market continues to expand, the major resin suppliers are beginning to apply more research and development effort to the production of resins specifically suited for such applications. At present no single resin type or product has dominance of the market. Inasmuch as an average formula may contain from 10-25% binder, the market for resin is attractive.

Aqueous coatings have been the subject of much R & D effort, but currently the performance of commercial solvent based formulas is not matched by aqueous systems. Economics is the prime mover in these developmental activities. For those companies already well equipped with water-based coating hardware it is an extremely attractive proposition. For those producers equipped with solvent coating facilities the urge to develop water-based systems is not as strong. Depending on the solids content of the formulation and the speed capability of the coating machine, it can be shown that only small differences in the cost of off-machine solvent-based zinc oxide coatings and aqueous systems exist. The present total market for such coated papers could easily be satisfied by a few of the large coating installations operated by any of the major paper coaters. Because the marketing of these papers is still largely controlled by the systems supplier rather than the independent paper producer and because the total market does not justify such production capabilities, it is doubtful whether much aqueous based production will appear in the near future. However, should an aqueous system be developed that has functional advantages in terms of either weight or performance, then of course the situation will change.

Dye Sensitization

The choice of suitable dye sensitizer systems for zinc oxide coatings is limited by several important considerations. The dyes must be selected and mixed to have maximum absorbtion in the visible spectrum and should match as closely as possible the output of the light source in the particular machine for which the paper is being optimized. In addition to this most important requirement the dried coating should be a neutral grey shade with as high a conventional "brightness" as possible. If one examines a series of commercial coated papers next to one another the variations in actual hue become quite apparent. Commercial products tend to cover the shade spectrum from pink to blue.

Specific functional dyes are described in the patent literature. Among the more common dyes that have found use are rose bengal, eosin, fluorescein, bromphenol green, and auramine. Each specific resin/oxide system must be matched with a suitable dye combination that is soluble in the specific resin system employed.

Solvent Selection

Specific solvents such as toluene, xylene, methyl ethyl ketone, and V.M.&.P. naphtha are used. In some cases solvent mixtures are employed to produce the desired mix viscosity for the specific coating equipment involved. The dyestuffs themselves are often added from methanol solutions and, depending upon the concentration of dye, more or less methanol is automatically added to the slurry. The methanol also tends to lower viscosities in most systems and can be an effective control technique to adjust for small variations in resin.

Coating Equipment

Suitable coaters for *Electrofax* applications include roll coaters, rod coaters, and some knife coaters. In the majority of instances reverse roll coating equipment is chosen. This equipment is capable of applying uniform thicknesses of wet coating and can operate at speeds up to 500 ft/min before mechanical problems of coating "throwout" are encountered.

Dry coat weights are in the range of 15-30 lb/3000 ft^2. Each formula/ paper combination tends to have a very real lower limit beyond which imaging performance is seriously affected. Most systems show improvements in performance as a function of applied coat weight and some papers have appeared in the market place with coat weights of over 30 lb. However, the trend is definitely in the opposite direction. Economic factors are becoming increasingly important as prices tend to erode under intense competition.

Hot air drying is normally used for these systems. It is necessary to exercise close control of the drying conditions because excessive heat can sometimes cause adverse side reactions in the formulation itself and occasionally problems in the performance of the base stock as well. Excessive dryness in the finished paper tends to promote higher resistivity than is good for the performance of the system in certain equipment and in the low ambient humidities normally encountered in midwinter in many offices. In addition to these problems, too rapid removal of solvent from the coating can produce excessive pinholing, which also reduces the overall quality of copies made on the paper. Large volumes of medium velocity air at low temperatures generally constitute the best initial approach to drying of these papers. If the specific resin system being employed requires a curing by heat to generate maximum hardness in the finished coating, this can be done in a later zone after the solvent has been removed.

Because coatings that contain zinc oxide mar easily when rubbed by metal objects, only turning rolls should be in contact with the coated surface.

Pigments and Resins

Other pigmentary photoconductors have been suggested for use in these systems. U.S. Patent 3,121,006 (issued to A. E. Middleton and D. C. Reynolds and assigned to the Xerox Corporation) claims a great number of useful resin/pigment systems for the production of so-called xerographic plates. This terminology is used in the broadest sense to include all photoconductive materials dispersed in insulating resin binders coated on a substrate. For example, the patent lists and claims as useful photoconductive insulators the following materials:

Zinc sulfide
Cadmium sulfide
Zinc selenide
Cadmium selenide
Titanium dioxide
Zinc/cadmium sulfide phosphors
Arsenic tri-selenide
Lead oxide (Pb_3O_4)
Mercuric oxide
Indium trisulfide
Arsenic sulfide

Among useful resin binders specified in the same patent are the following:

Polystyrenes
Silicone resins
Acrylic and methacrylic esters
Chlorinated rubber
Vinyl polymers and copolymers
Cellulose esters and ethers
Alkyd resins

In addition to the individual resins specified the patent suggests the use of mixed resin systems with or without the use of certain plasticizers to improve functional properties such as adhesion, flexibility, and blocking resistance. Suitable plasticizers suggested are the phthalates, phosphates, and adipates. In general, Middleton's specifications indicate a preference for binders having relatively high polarity groupings such as carboxyl, chloride, and acetate. He states that charge injection from the photoconductor to the binder is improved by the presence of such groupings.

Perhaps the most important conclusion to be drawn from this patent and others dealing with the same subject is the wide choice of materials available to the formulator. This broad flexibility has led to the development of many individual resin/pigment systems that have good functionality in a wide cross section of commercial copy machines.

Because of the commodity price structure and inherent pigment properties of zinc oxide, it is the normal pigment of choice for present commercial systems. Although dye sensitization does lower the overall whiteness of the coatings, they are still markedly superior in appearance to many of the photoconductors listed by Middleton that have inherently strong color. In the case of a transfer xerographic system wherein the photoconductor is constantly recycled, color is not a factor because the final copy is delivered on conventional bond type papers.

Organic photoconductors are being studied in many laboratories for their possible application in these fields, and one offset plate available on the market today is coated with such a material. It is expected that major breakthroughs in materials and applications will come about in the near future.

Market Data

The introduction of commercial systems for *Electrofax* copying dates back to 1960-61. By 1965 the consumption of coated papers had reached approximately 10,000 tons. Current estimates indicate that more than 25,000 tons will be produced in 1966. At the projected growth rate of 20% per year over the near future, paper usage is likely to reach nearly 50,000 tons a year, at which time a leveling tendency can be expected. The successor system to *Electrofax* has not yet appeared but considering the diminishing time interval between each major new system introduction it would appear reasonable to anticipate perhaps a ten-year cycle for this system to arrive at maturity and subsequent obsolescence. Although diffusion transfer systems are considered to be already obsolete by some, a market for papers and some machines, albeit a declining one, continues to exist. Thermal copying also has reached a plateau with some indication of decline but recently introduced new variations of thermal processes show promise of markedly improving the situation.

The need to copy seems to increase almost in a geometric ratio with the introduction of newer and simpler ways of copying.

Future Developments

Direct electrography is being intensively studied with the aim of developing printout systems capable of very high-speed operation. These systems apply the image directly in the form of an electrostatic charge pattern and do not require photoconductive coatings. They do require a dielectric surface of high resistivity to capture and hold the applied charge until it is developed with a toner of some sort. Commercial equipment already exists in limited applications for such purposes as high-speed label printing via tape inputs and for transmission of documents and messages to remote locations from a central transmitter.

Color electrophotography will continue to be developed and at present systems are being offered for color-proofing of color separations ahead of the press. The development of good reversal toners could possibly lead to the introduction of amateur systems for negative printing purposes.

Perhaps the largest potential market for such specialty papers is yet to come. This is in the area of information retrieval systems. Much has been written about the knowledge explosions and publication explosions that have taken place over the past twenty years. Development engineers are hard at work to produce practical information storage systems that can feed out printed copies of all sorts of information on demand from computer memories and microfilm memory systems. Paperwork continues to expand in response to the inexorable demands of Parkinson's Law, not to mention real needs of business and technology, and although electronic data processing will greatly help to control the projected mountains of paper, it is likely that hard copy demands of all types will continue to grow. The specialty papermaker and coater has every reason to anticipate a healthy future if he maintains contact with developments in this dynamic and expanding technology.

REFERENCE

1. Vaurio, F. and Fird, D.P., *TAPPI* **47**, No. 12, 163A (1964).

BIBLIOGRAPHY

General
Barnett, B. H., "Market for Office Copy Paper," *TAPPI* **48**, 79A-82A, Nov. 1965.
Anon., "Cites Growth of Copy Industry," *Office Appliances*, p. 134 (Sept. 1964).
Hoddeson, D., "New Look in Copying: Heightened Competition is Intensifying Both Rewards and Risks," *Barron's*, p. 3, (Nov. 23, 1964).
Holmes, D. C., "New Copying Methods Can Be Expected," *Office*, p. 162, Jan. 1966.
Anon., "Revolution in Office Copying," *Chem. & Eng. News*, 114-125 (July 13, July 20, 1964).
Worthington, R. L., "Skyrocket Rise of Reproduction," *Office*, p. 107, Jan. 1964.
Anon., "Annual Statistical Report," *Reproduction Methods*, No. 2 (Feb. 1966).

Diazo Systems
Weiner, J. and Roth, L., "Diazotype Papers-Bibliographic Series No. 220," The Institute of Paper Chemistry, Appleton, Wisconsin (1965). Sponsored by the Sensitized Papers Section of the TAPPI Coating Committee.
Kosar, J., *Light Sensitive Systems: Chemistry and Application of Nonsilver Halide*

Photographic Processes, J. Wiley & Sons, New York, pp. 194-320.

Habib, D. P., & Hodgins, G. R., "The Diazotype Process," paper presented at the SPSE Symposium on Unconventional Photographic Systems, Wash. D.C., Oct. 29-31, 1964.

Glossary, National Association of Blueprint and Diazo Coaters, (1956).

Muller, P., "Precoating of Diazotype Paper," preprints of 16th Tappi Coating Conference, Portland Oregon, May 9-13, (1965).

Diffusion and Dye Transfer

Anon., "The Revolution in Office Copying," *Ibid.*

Anon, "Die Blitzcopie," *Foto-Kino-Technik*, 4, 25-26, 1950.

Anon., "Chemicals Find New Office Copying Jobs," *Chem. Week*, p. 31 (Nov. 12, 1960).

Yackel, El C., U.S. Patent 2,596,754, Photomechanical Copy Method, May 13, 1952.

Yutzy, H. C. et al., U.S. Patent 2,596,756, Photomechanical Copy Method, May 13, 1952.

Yutzy, H. C. et al., U.S. Patent 2,739,890, Photographic Reproduction Process, March 27, 1956.

Yutzy, H.C. et al., U.S. Patent 2,716,059, Photographic Transfer Process, Aug. 23, 1955.

Yutzy, H. C. et al., U.S. Patent 2,675,313, Photographic Reproduction Process, April 13, 1954.

Haist, G. M., et al., U.S. Patent 2,875,048, Combined Photographic Developing and Stabilizing Solution, Feb. 24, 1959.

Lodewijk, J., et al., U.S. Patent 3,203,796, Use of Starch Ether Layers in Diffusion Transfer Materials, Aug. 31, 1965.

Wendt, R. U.S. Patent 3,211,551, Diffusion Transfer Coating, Oct. 12, 1965.

Belgian Patent 660,190, Receptor Sheet for Diffusion Transfer Reversal Process, June 16, 1965.

Przezdziecki, W.M., Brit. Patent 1,010,202, Receivers for Diffusion Transfer, Nov. 17, 1965.

A. B. Dick Co., Neth. Appl. 6,500,536, Silver Halide Diffusion Process Paper With Silica, July 19, 1965.

Mead Corp., Neth. Appl. 6,500,727, Silver Halide Diffusion Process Paper With Calcium Carbonate, July 21, 1965.

Liebe, W., et al. Belg. Patent 622,297, Apparatus for Dry Copies by the Silver Diffusion Process, March 11, 1963.

Buskes, W., U.S. Patent 3,091,528, Light Sensitive Sheets for Making Positive or Negative Pigment Transfer Images, May 28, 1963.

Liniberger, W., U.S. Patent 3,103,153, Developing Device for Making Copies by the Diffusion Transfer Process, Sept. 10, 1963.

Land, E., U.S. Patent 3,113,866, Emulsions or Dispersions Useful in Silver-Transfer Reversal Photographic Process, Dec. 10, 1963.

Thermography

Roth, L., and Weiner, J., "Thermographic Papers," *Bibliographic Series No. 221*, Institute of Paper Chemistry, Appleton, Wis. (1965). Sponsored by the Sensitized Papers Section, TAPPI Coating Committee.

Gold, R., "Thermography-State-of-the-Art Review," preprints of the SPSE Symposium on Unconventional Photographic Systems, Washington, D.C., Oct. 29-31, 1964.

Anon., "The Revolution in Office Copying," *Ibid.*

Hoshino, S., et al., "A New Thermographic Process," paper presented at the SPSE Symposium on Unconventional Photographic Systems, Oct. 29-31, 1964.
Patents—Excellent thermographic patent reviews are presented by Gold, R., *loc. cit.* and Roth and Weiner, *loc. cit.*, and they should be consulted by the serious reader interested in tracing the development of thermography from both the paper and machine sides. The following U.S. Patents are selected from the above sources and are presented as a guide only:

Machine Patents:	2,630,484, 2,740,895, 2,891,165,
	2,927,210, 3,053,175, 3,056,904,
	3,128,379
Paper Patents:	2,710,263, 2,859,351, 2,880,110,
	2,927,039, 3,020,172, 1,939,232,
	2,129,242, 2,630,484, 2,663,657,
	2,675,332, 2,855,266, 2,899,334

Electrophotography

Mott, G. R., "Electrostatic Electrophotography 1964," preprints of Symposium on Unconventional Photographic Systems, Oct. 29-31, 1964, Washington, D.C.
Anon., "The Revolution in Office Copying," *Ibid.*
*Weiner, J., and Roth, L., "Electrostatic Printing," Bibliographic Series 212 with Supplement No. 1, Institute of Paper Chemistry, Appleton, Wisconsin, 1965. Sponsored by the Sensitized Papers Section of the TAPPI Coating Committee.
Young, C. J., and Greig, H. G., "Electrofax, Direct Electrophotographic Printing on Paper," *RCA Review*, **XV**, No. 4, Dec. 1954.
*Schaffert, R. M., *Electrophotography*, Focal Press, New York (1965).
*Dessauer, J. H. and Clark, H. E., *Xerography and Related Processes*, Focal Press, New York (1965).
Anon., "1966 Annual Statistical Report," Reproduction Methods, No. 2 (Feb. 1966).
Uhl, W. F., "Base Paper-Barrier Properties Relating to the Electrofax Process," preprints of the 16th TAPPI Coating Conference, May 9-13, Portland, Oregon (1965).
Diamond, A. S. *et al.*, "Some Considerations in Paper Substrate Selection for Electrophotographic Coating," preprints of 16th TAPPI Coating Conference, Portland, Oregon, May 9-13, 1965.
Langston, D. J., "Dye Sensitized Electrophotographic Paper: Pre-Exposure Effects," preprints 16th TAPPI Coating Conference, Portland, Oregon, May 9-13, 1965.
Heidecker, S. C., and Taft, D. D., "Electrophotographic Paper Binders," preprints of 17th TAPPI Coating Conference, May 9-12, 1966, Chicago, Ill.
Vaurio, F. & Fird, D. P., "Electrically Conductive Paper for Nonimpact Printing," *TAPPI* **47**, No. 12, 163A-166A, Dec. 1964.

---

* Rather extensive patent reviews may be found in the cited references. Schaffert lists the patents of special relevance along with a very brief abstract. The Weiner and Roth bibliography also annotates the patent references.

chapter 7

Cover Papers, Synthetic Leathers, Coated Tag, Boards, and Photo Mounts

R. H. MOSHER

Cover papers, imitation leathers, and various types of coated-tag stock and boards are the heavier weight products of the paper-converting industry. They are manufactured from heavy-weight base paper or lined board or from stock made by laminating together several plies of boards or paper. The final product may be coated or printed on one or both sides, or the individual facing sheets may be coated or printed before laminating. They may possess only decorative characteristics, or may also be functional in its applications.

The products are grouped into four major classes:
1. Cover papers.
2. Synthetic or imitation leathers.
3. Coated tag and boards.
4. Photo mounts and special mounting boards.

Because the four types are so different in their characteristics they will be discussed separately.

COVER PAPERS

Cover papers can be divided into two main types: uncoated covers and specialty coated or printed covers. The uncoated types are generally given only a fleeting treatment by converters, but the coated or printed covers undergo several converting operations before they are ready for use.

Uncoated Covers

The uncoated grades are generally converted through the application of an embossing pattern or printed design on one or both sides of the sheet. The most common grades of this type are made from a colored base paper. Some of the most important characteristics of the sheet are good folding strength and tear resistance, good resistance to fading, and good wet and dry abrasion resistance. Such papers have been print-embossed or spanished to obtain special effects.

Coated Specialty Covers

These are usually made from a strong, long-fibered base paper, such as a kraft or sulfite sheet. The final sheet may be a single ply or a laminated structure, depending upon the desired weight and method of making the final product and may be white, tan, or a color complementary to the coating. The sheet may be coated on one or both sides, coated and waterproofed on one or both sides, embossed on one or both sides, or printed, print embossed, topped, or spanished or two-toned, depending upon the requirements.

Coating

The base sheet may be coated with a pigmented casein, glue, or other protein coating, or with a pigmented synthetic resin (plastic) coating laid down by a solvent, emulsion, hot-melt, or organosol technique. Where a two-side-coated product is required, two such sheets can each be coated on one side and then laminated back to back to obtain the desired two-side-coated product. A heavy weight single sheet can also be coated on both sides. The protein-based coatings can be formulated to possess good flexibility and dry-abrasion resistance, but their wet-abrasion characteristics, even when waterproofed, are generally not completely satisfactory. The plastic-coated sheets generally possess good flexibility and good wet- and dry-abrasion resistance. The use of the newer synthetic resin latices together with the protein adhesives and the new water-solubilized synthetic resin-base adhesives will no doubt improve the results with water-base coatings. The coatings are pigmented to produce the required colors and the proper choice of pigments will result in the desired nonbleeding and light-fast characteristics.

Waterproofing

Where a nonwaterproof pigmented coating is used to give the required color and surface characteristics to the sheet, a clear, waterproof top coating or sizing is often applied. Such a sizing can be formulated with waterproofed casein or other protein or with one of the new water-soluble synthetic resins that can be insolubilized after application to the sheet. A synthetic resin (plastic) coating laid down by a solvent, emulsion, hot-melt, or organosol technique also offers distinct advantages.

The waterproofing, or top-sizing operation, produces a durable, nonsoiling surface that withstands handling and general abuse, fingermarking or scratching, as well as moisture and solvents. The base sheet can be coated on two sides or two single coated sheets can be laminated to produce the desired effect. When two waterproofed sheets are to be combined, however, the problem of water or solvent removal from the glue line may be important, and the use of special techniques involving coating and drying the adhesive before combining the sheets may be necessary.

Most specialty covers are embossed with leather designs, although various cloth and other special designs are also used. The embossed effects are usually obtained by one of three different methods.

Machine Embossing

The finished coated or uncoated web is run through an embossing machine, which is composed of two rolls, one of which is engraved with the desired design to produce the impression. The second, or supporting, roll acts as a backing for the sheet during the embossing operation and either a smooth or reverse embossed effect can be obtained on the back side of the sheet depending upon the type of backing roll.

The two rolls may both be steel-engraved and mate as they turn and in this case both sides of the sheet are embossed with the same pattern, the face design being depressed and the back design raised. An alternate procedure to obtain a similar effect is to use an engraved steel embossing roll and a paper-filled backing roll that are run in together under pressure without any paper in between until the steel roll has imprinted its design uniformly into the paper-filled roll. These two rolls can then be used to produce a two-side-embossed job similar to that obtained with matching steel rolls. When a smooth backing roll is used and only sufficient pressure to emboss the sheet, the face of the sheet takes the design and the back remains practically smooth.

Sheet Embossing

Some mills have equipment for embossing cover papers in sheets rather than in rolls and this is accomplished by using engraved plates mounted in embossing presses. The process is slow and expensive, and the sheet size that can be handled is very limited. The effect is not as pronounced as when the paper is web-embossed.

Plater Embossing

The third method of producing embossed paper is known as plater embossing because the effect is obtained by the use of plating calenders. The cover paper is placed between a pattern sheet and a smooth metal backing plate and several of the units are made up and run in under smooth heavy-pressure application rolls. This kind of embossing is usually confined to cloth finishes where the actual cloth is used as a plate but a lumpy paper is often used to obtain a ripple finish.

Special effects can be obtained by printing the surface of the cover paper before or after coating and before embossing and particularly where the printed design is partially or completely synchronized with the embossing pattern.

Print-embossing equipment where the embossing roll carries an ink film and colors the embossed design with a contrasting or matching shade to produce a two-tone effect can be used to obtain interesting results. Another possibility is to apply a coating or a surface treatment during the coating operation that discolors on heating. The embossing roll is then heated so that simultaneously with the pressure high temperature is applied to the same area of the sheet. Such a sheet produces a unique two-tone effect with different degrees of shading where the pressure is varied by the depth of embossing. With such embossing rolls, a smooth backing roll is commonly used.

A spanished cover stock is obtained by embossing the sheet in the usual manner and then passing the embossed sheet through a coating unit where an excess of a contrasting or blending pigmented coating is applied, and then doctored off so that the applied coating only remains in the depressions. This technique can be used to produce some startling effects that cannot be obtained in any other may. A knife coater is usually employed to produce this effect.

A topped cover is the reverse of the print-embossed or spanished sheet because a smooth roll or etched roll carrying a thin film of ink or coating is run against the previously embossed sheet so that only the high points are colored. Here again very unusual effects can be obtained and similar, blending, or contrasting colors can be used.

Some specialty covers are made with a high-gloss surface finish. Such papers can be produced by laminating a sheet of high-gloss paper to a cover base sheet. Another possibility is to laminate a sheet of coated or colored cellophane or acetate film with a clear laminant or a clear film to the face of a cover base stock. By this means, a high-gloss sheet that possesses the flexibility and bulk needed in a cover can be obtained. It is difficult to polish heavy papers to produce glossy surfaces in the normal manner without adversely affecting other necessary properties of the sheet. Other special surface effects are attained by using mica or velour coatings.

The desirable properties of a cover paper are good folding and tearing strength, good fade, mold and fungus resistance, and a soil resistant surface. Grease resistance is important if the fabricated product is to be handled by greasy fingers. The abrasion resistance, both wet and dry, must be exceptional because the sheet is continually exposed to handling and scuffing. Printability, or the ability of the customer to print any desirable legendary or decorative characters on the surface of the sheet, is extremely important. The surface should also, if possible, take the usual printing inks, because the small printers, who as a group are large customers of the specialty converting industry, do not like to have to use special inks. The acceptance of gold leaf is also very desirable.

Cover papers are produced in a wide variety of basis weights, from the lightweight papers, known commonly as leatherettes, which are coated one side, up to double-weight, two-side-coated covers used on large brochures or manuals, and for the mechanical bindings that are becoming so popular.

The principal applications of cover papers are for displays, fancy box coverings, and wrap-around covers in the light weights or leatherettes. The medium- and heavy-weight products are used for catalog, booklet, folder, proposal, and instruction book covers; manual and reference book covers; and counter cards and other miscellaneous converted items. During World War II, various government agencies recognized the importance and necessity of protecting their specifications and other printed matter. Cover papers were widely used in such applications. Printers, advertising managers, advertising agencies, and buyers of printed matter have long realized the importance of protecting their printed matter and have been large users of such cover papers.

SYNTHETIC OR IMITATION LEATHERS

Considerable work has been done in recent years on papers made to have the appearance, properties, and feel of genuine leather. Such papers are specially treated so that exceptionally high strength is built into the base paper. They are coated to simulate the surface appearance and properties of real leather and are embossed to look like leather. The commonly used base papers are 10, 17, 20, 25, and 35 mils thick; they are made from kraft, rope, sulfite, alpha, and rag pulps.

The major requirements in the base stock are high tear and tensile strength, good flexibility, and high bulk for good embossability. Such properties can best be obtained by an increase in the internal bonding of fiber to fiber within the sheet while still maintaining low density and high bulk. The base paper is made from strong and long fibers, with an open structure, low density, and high caliper, and it is then saturated with a synthetic resin or natural or synthetic rubber latex. The rubber or resin particles possess the property of bonding the fibers together to produce the required physical strength while still retaining the open and bulky characteristics in the saturated sheet. High tensile and tearing strengths are obtained, and the sheet is flexible and bulky so that it will readily assume and retain the embossed pattern. The latex may be added to the sheet at the beater or on the paper machine, or it may be incorporated in a secondary converting operation.

The saturated paper is then coated with a synthetic-resin-base, pigmented, plastic-type coating. Solvent, dispersion, hot-melt, or organosol techniques can be used. The saturated paper makes a very poor coating base stock because of its high porosity, low density, and relatively rough surface because the sheet, either before or after saturating, cannot be calendered or smoothed to any extent without reducing the bulk. Multiple coating operations are usually required to produce the desired surface effect and as many as five separate coats of lacquer have been used.

Other techniques generally require fewer individual coating operations for an equivalent job, but water dispersions need at least three and organosols generally require at least two coats to obtain the desired surface effect. The final coating is grease- and chemical-resistant, ages satisfactorily, and resists scuffing and abrasion. By a proper choice of pigments, the desired colors, which possess light stability and fade resistance, can be obtained. If desirable, additives can be incorporated into the coating or base sheet to make the finished product smell like leather.

The coated sheet can then be embossed, print-embossed, spanished, topped, or otherwise treated in a manner similar to that for regular cover papers. On account of its resiliency and bulk, however, the saturated sheet takes a deep embossing pattern easily and the proper base stock, coating and embossing treatment produce a sheet that resembles real leather. The embossing patterns used are usually designed to simulate the real leather grain as closely as possible and special leathers are often used directly as the basis of roll design and reproduction.

These papers are used for the same purpose as the better grades of cover papers as well as for hat bands, belts, dress accessories, wallets, folders, parts for shoes, handbags, automobile accesories, chair and bar decorations and coverings, and luggage covering.

COATED TAG AND BOARDS

Coated tag and boards are heavy coated papers and are usually produced in converting plants that specialize in handling the heavier weight papers. These products are usually manufactured to specifications as to caliper and finish, and the finished sheets are mostly printed before they find their eventual use. The coatings are often applied to surfaces that are not as smooth and uniform as are common in lighter-weight base papers, and as a result the customary coating weights are heavier and may contain more adhesive than those used in the lighter-weight field.

The manufacturing problems are other than those with the lighter-weight papers in that there is less footage of paper on a roll than with lighter-weight papers, thus necessitating more frequent roll changes and handling; two-side coating is common; the heavier-weight coatings require more drying capacity; and it is difficult to handle these sheets, which are stiffer and heavier than the ordinary coated papers, on festoon lines without getting streak marks and cracked paper at the sticks.

A distinct problem arises in producing such coated papers to uniform caliper on large widths, because the raw stock itself may vary widely in thickness across the web. The coated rolls are usually calendered at least once and often twice to achieve the desired uniformity of caliper and to obtain the required surface finish.

Coated Tag

Coated tags are usually made from base papers manufactured on the fourdrinier machine, but many of the medium and heavy weights are made from cylinder stocks. The surface of the base sheet is usually rougher than is desirable, because of the thickness of the stock, and a heavy coating is necessary to level off the sheet surface and cover it sufficiently so that the base imperfections and sheet texture will not appear as a mottle after calendering. Such coating weights are also desirable to produce a satisfactory smooth printing surface and attain the specified sheet thickness. The base stock for converting purposes is usually fairly highly sized so as to hold up the applied coating and is white, buff, or manila in color. The thickness ranges from 6 to 20 points depending upon the use.

The requirements of the coated stock depend to a great extent upon the end use of the product and the specifications of the customer. The sheet should possess good folding and tearing characteristics and have a surface smooth enough to insure ready printing. The formulation of the coating compound depends upon the abrasion resistance requirements of the customer and the type of printing equipment available to produce the finished tags. A sheet that will be made into laundry tags and must pass through a laundering operation is different from a tag to be attached to a new shirt for identification or pricing purposes in a retail store. The printing requirements also vary as some tags are printed with surface drying inks by the rotogravure process whereas others are handled on the usual flat-bed presses with oxidizing oil inks. The amount of adhesive in the coating formulation may range from 18 to 30% and, in some cases, the coatings are waterproofed by the addition of shellac or formaldehyde. The pigment composition may vary from English clay or a carbonate pigment for good ink absorbency to a fine domestic clay or satin white to produce smoothness and a glossy surface.

The fiber formulation of the base stock can vary from rope, jute, sulfate or sulfite to soda or groundwood pulp, according to the use requirements of the final tags. Some tags must be strong and resistant to tearing, whereas others must be easily torn in half for identification or verification purposes.

The manufacturers of tabs have formed an association known as the Tag Manufacturers Institute. This group has drawn up a series of specifications for colors, fiber composition, and for the physical properties of the sheets they purchase for their manufacturing operations. The specifications cover 15 colors as well as gold and silver. These are known as T.M.I. shades and are usually specified on orders to the coating mill. In the specifications are also outlined in detail their test procedures and tolerances, as well as the terms and conditions of sale to the tag-consuming industries. Tag stock is supplied either in rolls or in sheets depending upon the requirements of the customer.

Tag stock in sheet form is usually termed tough check. This product is usually made from a single-ply sheet, containing long, tough fibers and is coated on one or both sides in white or colors. Tough check is usually supplied in three-, four-, six- and eight-ply designations, which refer to 12/1000-, 18/1000-, 24/1000-, and 30/1000-in. thicknesses. The standard stock is usually 22 × 28 in. in size and packed in 100-sheet packages.

Tag stock is used for making all kinds of tags, file folders, heavy envelopes, score cards, menus, tickets, identification checks, car signs, and in applications where its strength and rigidity are desirable.

COATED TAG AND BOARDS

Coated boards include those products made from base sheets manufactured on the cylinder machines and are 12/1000 in. or more in thickness. Coated boards, coated and specialty blanks, and display boards also belong to this group. The coated boards are usually made up in thicknesses ranging from 12 to 28 points and are based on a single- or multiple-ply base stock. The blanks and display boards are manufactured in thicknesses ranging from 12/1000 to 78/1000 in. and are also made from single- or multiple-ply base sheets.

Coated boards are generally made by coating with clay or other pigment and casein directly on to the board surface. The base stock may be made from waste paper and bleached or unbleached kraft or sulfite pulp and is generally lined on one or both sides. A fairly heavy coating load, 10 to 24 lb (23 × 29—500), is applied because the board surface is often relatively rough and may contain coarse matter or dirt that must be covered up to make a good appearance and printing surface. The coated sheet is always supercalendered to smooth out the surface and attain the desired caliper, and care must be exercised not to emphasize any imperfections in the base stock that may show through the coating.

The coated blanks are made from a cylinder-lined or multiple-ply pasted base sheet, which results in good durability and a certain rigidity not present in the average coated board. The minimum-thickness blank is manufactured from a two-ply, cylinder-lined board, but the heavier weights, usually in excess of six ply, are made from a laminated structure composed of a middle and two liners or facings. The middles are usually made from waste paper, either plain or deinked; groundwood; bleached or unbleached kraft or sulfite; or some mixture of these pulps. The facings are either kraft or sulfite base and may be white or colored, depending upon the use requirements of the stock. There are several classes of blanks:

a) *Uncoated blanks*—middle of news or groundwood and bleached kraft or sulfite facings in either white or colors.

b) *Coated blanks*—single or multiple coated on both sides and calendered to a high finish. The coating is usually white and shows a high whiteness, usually on the blue-white cast.

c) *Coated railroad*—a lined news middle with a two-side coating, usually in colors.

These blanks are almost always coated with a casein-base formulation pigmented to obtain the desired color shade. The coatings are fairly highly sized, but generally not to the extent of a tag stock, the top sizing content usually being in the neighborhood of 24%. The sheet is characterized by a certain stiffness and rigidity in the pasted weights, because of the multiple-ply structure made up by a starch or glue pasting operation. This property is particularly desirable in the blanks as their biggest use is in the display and mounting field where they must hold up without sagging.

The usual sizes of display blanks range from two-ply or 12/1000 in. in thickness to twenty-four-ply or 78/1000 in. This stock is commonly sold in sheets packed 25 to 100 to a package, depending upon the sheet thickness, in standard sizes of 22 × 28 in. or 28 × 44 in. The grain direction in these sheets is usually specified since it is important that in a show card or display card the grain should be perpendicular to the base. Where the rigidity is not important, but the folding qualities are desirable, the grain should be specified so that the fold runs in the grain direction.

The principal uses for these coated blanks and boards are for display cards, calendar backs, menus, car cards, tickets and checks, show cards and mailing cards.

PHOTOMOUNTS AND SPECIAL MOUNTING BOARDS

Photomounts and other mounting boards are used to hold, mount, and frame photographs, calendars, and advertising matter for display purposes. The stock is usually made by pasting two facing papers on a news middle. The characteristics required in the sheet are good rigidity, freedom from curl and warp, good folding qualities, and resistance to fading. The rigidity is needed because folders, mounts, or displays are usually set on edge on a table or counter and it is necessary that the board does not bend or flex. The second major requirement is flexibility as photomounts must be opened and closed many times without cracks developing on the folds.

Mounting boards are made in three different grades:

Soft Folder

Soft folder stock is made from special grades of long-fibered pulp that produce a clean flexible sheet; sulfite, bleached kraft, or rag pulps are utilized. The stock is usually made up to 25 points in thickness, if a single sheet is required, but where the usual heavier, more rigid structures are needed, it is customary to laminate two or more plies to obtain the desired thickness. Flexible laminants produce the best sheet, and the multiple-ply structure can be made warp- and curl-free, yet to possess the desired rigidity. The sheet can be printed or embossed if either is desirable.

Cylinder Mounting Boards

Mounting boards are manufactured on a cylinder machine and are lined on one or both faces according to the requirements for the sheet. Such a board can be used for many applications where the use requirements are not too difficult to fulfill. The liners are generally a sulfite or bleached kraft, white or colored, and the centers are often groundwood, reused news stock, or chipboard. If desirable, the sheet can be printed or embossed. Such boards are often printed with an over-all design simulating an embossing or other special effect for particular applications.

Pasted Specialty Mounts

Pasted mounts represent the largest field of application and are generally produced in either two- or three-ply construction, depending upon the thickness required and the end use of the sheet. A three-ply stock is produced by combining two plies of facing paper to a middle or center. The thickness of each ply is determined by the desired thickness of the finished board. The final thickness is also related to the rigidity and also the flexibility of the board. The proper adhesives must be used and the stock must be properly dried and seasoned to avoid any curling or warping in the final product. The sheet is usually a balanced structure and the problem of warp or curl is thus minimized, particularly on heavier mounts. Proper seasoning is good insurance against curl and warp. Most mounting boards are held under standardized conditions for at least 3 days, and usually longer, before shipments to the consumer.

The two-ply mounts are generally made with one or both faces coated, printed or embossed, and after the sheet has been pasted, an overall embossing pattern is sometimes applied. Such a sheet, if made properly, has a balanced structure, is flexible, and can be fabricated readily; yet because of its laminated structure, it still has a certain amount of rigidity, even in the lighter weights.

The three-ply structure is more difficult to engineer and to produce. A news middle, which is a cylinder sheet composed of pulp made from old newspapers, is generally used to supply the desired bulk and thickness to the finished mount, but because the folding and scoring properties of such a sheet leave something to be desired, the two facing papers must have extremely good folding and flexibility characteristics. Kraft facings are generally used and if laminated properly the finished mount will be flexible so that the photo mounts can be opened and shut many times without cracking at the hinge, yet the entire structure will be rigid and stand without warp or curl on a table top. A secondary, yet important, problem with the middles is their thickness. Because thickness is directly related to the mount rigidity, the larger the expected photograph or easel size, the thicker and heavier must be the middle used in the pasting or laminating operation.

Mounting boards are generally made with one side coated or printed and the reverse side with a plain or colored uncoated facing when they are used for display work. The photo mounts are generally made with both faces coated or printed as they are handled and viewed from both sides. The coatings may be pigmented casein or other protein or plastic resin-base types, and the nonplastic types are generally top-sized or coated with a clear, protective, waterproof and greaseproof coating film to give the desired scuff and abrasion resistance. A heat-sensitive top coating that discolors under heat and pressure is often used so that special two-tone embossed or die-stamped patterns can be applied by the customer where this is desirable. Fine embossing patterns are often used to give the outside cover of the photomount a special finish. The inside face or insert that holds the photograph or other decorative matter is generally white, buff, light gray, or cream-colored with a smooth or lightly embossed surface finish to provide the most suitable background for the photograph, calendar, diploma, or other decorative or legendary matter.

It is also common practice to print an overall design on the outside and inside of some less expensive mounts, a dark gray and white on the outside and a light gray and white on the inside, which will blend with the black-and-white photograph. This is usually an unobtrusive shadow pattern that tends to make a more attractive appearance and covers up fingermarks or other soiled spots that occur to deface the mounting after long use. Such printed patterns may or may not be top-sized with a waterproof and greaseproof coating after printing.

The design and over-all coatings must be fade-resistant and the proper choice of inks and pigments is very important. As one side is often embossed whereas the other side has a smooth, glossy or matte finish, the two facing sheets are usually coated or printed separately and then laminated to produce the finished sheet.

Display boards can also be built up with a foil or glossy film facing, and sometimes even with special phosphorescent or fluorescent coatings, where specific effects are desired. Such sheets are true specialties, however, and do not enter the general lines of most converters.

The main application for these products is photo mountings and display cards, but there are some more specific uses such as mountings for blueprints, charts, and graphs; greeting cards; diplomas; announcements; and the many miscellaneous items that can be displayed in the windows, showcases, or counters of retail and wholesale stores and business establishments.

chapter 8

Asphalt and Polyolefin
Waterproof Papers

D. F. KRAMER and R. H. REEVES

Until about 1950 the paper industry generally accepted the terms "Asphalt Papers" and "Waterproof Papers" as synonymous, meaning papers saturated, coated, or laminated with asphalt to make them water resistant, waterproof, or water-vapor proof. Since that time, however, polyolefins and other polymers have found increasing use in the production of waterproof papers and the waterproof paper industry is going through a major transition in product composition. In further clarification, the above terms have been used generally to describe only grades utilized in industrial fabrication, building construction, and industrial types of packaging and not for packaging that may indeed be water-vaporproof but that utilizes fine papers or types of papers used in the food packaging industry. Only the industrial types of waterproof papers are described here.

ASPHALT WATERPROOF PAPERS

Within the scope of this section, one cannot outline completely all the grades of paper made by the asphalt-paper industry. A summary of many of the grades used for wrapping and packaging can be found under Federal Specifications[1] or listed in wartime packaging manuals, such as the General Army and Navy Specifications for Packing and Packaging of Overseas Shipments[2]. The simplest product produced by the industry is one consisting of a single sheet infused or saturated with asphalt. Such single-ply infused grades are considered usually as water-repellent or resistant, rather than waterproof or water-vapor proof. The next grade of asphalt-treated papers consists of those

192

in which two or more sheets of paper are laminated with a discrete layer of asphalt as the combining medium between each pair of sheets. These combinations may utilize a number of grades of paper that may or that may not be infused with asphalt before laminating. Composite grades also may employ a variety of reinforcing materials usually embedded in the asphalt in the form of prewoven scrim, or mechanically woven into the combination. Also included in the laminated grades are those combinations in which metal foil such as aluminum, copper, lead, or steel is combined with paper either as a facing or as an intermediate layer. In a third group of products a tack-free asphaltic coating is applied to one or both sides of a sheet. The base to which such coatings is applied may be a single sheet of untreated paper or it may consist of various combinations of asphalt infused, laminated, or reinforced laminated papers.

All these papers can be produced in a number of variations by using papermachine-creped base stocks so as to increase the stretch of the finished laminate, or they can be creped during or after the laminating operation to produce higher ratios of stretch in the finished product. Numerous other combinations of creping and embossing or corrugating to produce grades with stretch in both machine and cross machine directions are available.

History and Present Importance of Asphalt Papers

Abraham,[3] in his comprehensive work on asphalt, gives a number of references marking the beginning of the asphalt-paper industry. As early as 1790, composition roofing was made in a crude manner by coating heated wood tar on plain paper placed over rough boards, but not until 1820 was a factory set up to impregnate paper with asphalt to produce a water-repellent grade for the manufacture of tarpaulins and protective packaging materials. In 1845 a British patent was issued to Williams[4] covering the method of producing asphalt-laminated paper and the product itself. In 1887 a U. S. patent was issued to Childs[5] on the same subject. The first patent[6] on the creping of asphalt papers was not issued until 1901, and the first patent on the reinforcing of asphalt paper was issued in 1907.[7]

Little reliable information is available as to the early volume of tonnage in the waterproof paper industry, but by 1936 production was estimated for that year to be 66,000 tons. Trade association records[8] indicate that the industry increased its output annually by 19 % until the pressure of war use swelled that tonnage to a record high of

362,000 tons in 1944. Total production of the waterproof industry fell to 191,000 tons in 1947 and remained at this level until 1957, at which time production began to decline to 130,000 tons in 1963. Although it would appear from this that tonnage of "waterproof" papers since 1956 is shrinking at a rate of $3\frac{1}{2}\%$ annually, there also has been a decline in membership of the trade association, reducing its reported annual production figures. The situation can be qualified further by noting a trend in packaging toward lighter basis weight polyolefin coated grades, so that the present yardage production of the industry probably has not decreased in proportion to the association's tonnage figures for "asphalt" papers.

Raw Materials:

Base Papers

The terms "base papers" or "dry sheets" designate the papers used for asphalt conversion. The characteristics of the base papers may vary widely, depending upon the grades being produced. Pulps for some low-grade sheets—for example, competitively priced sheathing paper—may utilize waste paper, Asplund fiber, or even groundwood. More permanent base papers (and, in general, for a large percentage of the total production of the waterproof-paper industry) are made only with 100% new kraft pulp. Papers that are to be infused or saturated with asphalt are usually of higher bulk, more porous and absorbent, and of lower bursting strength than those that are to be laminated. In turn, those that are to receive high-gloss coatings are produced denser and less porous than the intermediate laminating grades. Table 8-1 illustrates the base paper specifications. The range of tests listed in this table is only for 30 lb basis weight (24×36—500 sheets), but they serve to illustrate the relative range for various types of conversion. The asphalt-paper industry normally converts base sheets ranging from 25 lb basis weight to 90 lb basis weight (24×36—500 sheets), although both lighter and heavier grades are converted for various applications.

Laminating Materials

Asphalt derived from the refining of petroleum is used extensively in the waterproof-paper industry. Some producers of waterproof papers purchase asphalts and blend them with or without the addition of fillers. In general, asphalt is purchased on specifications and

TABLE 8-1

Typical Range of Tests on One Weight of Paper for Various Asphalt Conversions

Basis weight, 30 lb	Asphalting class	% Burst	% Tear	Caliper, in.	Gurley density	TAPPI water Penetration	% Moisture	Finish
	Infusing grades	75	200	0.0032	6	25	4.0	low
	Bag, sack, and container laminating grades	80	195	0.0031	13	30	4.5	medium
	Standard laminating grades	85	190	0.0030	18	30	4.5	medium
	Specialty laminating and gloss coating grades	85	180	0.0029	22	40	5.0	high

Basis weight: 24 × 36—500 count.
Burst in pounds expressed as percentage of basis weight.
Tear average in grams for sixteen sheets expressed as percentage of basis weight.
Gurley density as seconds for passage of 100 ml of air.
TAPPI procedure T-443m reported in seconds.

is used as received from the producer. All the major oil companies in the United States produce one or more grades for some segment of the waterproof-paper industry, and it is difficult to define rigidly what asphalt grades are to be specified as to working temperature and viscosity of the asphalt because of the wide variety of treatments represented by infusing, laminating, and coating, as well as the types of equipment utilized. Asphalts are purchased on the basic specifications of ring and ball softening point[9] and penetration[10], at a given range of temperatures. These two specifications fix approximately the flow point of the asphalt, its relative fluidity, and give some indication as to low temperature embrittlement. More comprehensive specifications quite frequently include requirements as to Furol viscosity, staining or oil migration tendency, odor, low temperature cracking, and loss of volatiles at specific temperatures. An approximate range of asphalts utilized by the asphalt-paper industry is shown in Table 8-2.

Increasing amounts of natural asphalts, coal tar and fatty acid pitches, wood and petroleum based resins, microcrystalline and paraffin waxes, and a wide variety of natural and synthetic polymers are also used. Although their application is limited to the blending of asphaltic materials for the roofing and shingle industry, the newer materials find broad application as components of nonasphaltic emulsion and hot melt formulations for those segments of the waterproof-paper industry that produce coated sheets and laminates. Nonasphaltic laminants and coatings can be formulated to give special properties to a combination such as resistance to bleeding at elevated temperature or pressure, inertness to chemicals such as wood sealers or slushing oils, good flexibility over a broad temperature range, and good moisture resistance after creasing.

TABLE 8-2
Approximate Range of Tests on Grades of Asphalt Generally
Used by the Asphalt-Paper Industry*

Application	ASTM D36-26, ring and ball softening point, °F	ASTM D5-25, penetration test, 77°F, 100g load, 5 sec
Infusing and saturating	60—120	Above testing range
Laminating	140—190	20—50
Coating	200—250	5—15

* These test ranges are broad and specifications for any one application must narrow the limits shown to a considerable extent.

Reinforcements and Other Materials.

Many materials are utilized for reinforcing certain types of asphalt laminated sheets although the principal material is glass fiber. Glass filaments are treated with binders and twisted into cords with tensile strengths in a normal range of 3 to 12 lb. Strands may be fed into the nip to produce a longitudinal one-way pattern, or woven into diamond-like two-way patterns by means of special attachments on the machine. A three-way pattern is made simply by running both one-way and two-way patterns simultaneously. In the absence of weaving equipment, prewoven scrim can be purchased in a rectangular weave in meshes ranging from ¼ in. to 2 in.

Sisal, jute, hemp, and cotton strands, once commonly used, have lost favor because of the superior properties of glass. Other synthetic fibers, such as rayon, nylon, or polypropylene yarns are used sometimes for special applications. In the past, a wide variety of burlaps was utilized, particularly in heavy-duty grades, such as for the wrapping of coiled steel. Since World War II little burlap has been available from producers in India, and therefore lightweight combinations of sufficient strength were developed as substitutes. Some highly specialized reinforced sheets have been produced with reinforcements consisting of plastic monofilaments, steel wire and, more recently, grades reinforced with steel strap[11]. Some producers manufacture combinations of paper and asphalt with metal foils, particularly aluminum foil of 0.0003-0.0005 in. thick.

Shipping and Application Equipment

Asphalt is received in paper cartons, steel drums, or tank cars; the last is the most economical and the one generally used by the large converters. If received in paper cartons, the containers are hand-stripped from the contents and the asphalt thrown into a melter in the form of large lumps. Steel drum shipments often are stripped by rolling off the steel drum after slitting, although in other cases the drums are emptied by inverting the open drums over receiving tanks and heating by means of canopies, steam coils, or by placing on heated racks until the contents flow out of the drums. Asphalt tanks are equipped with steam-heated coils. The cars leave the producer's plant at temperatures ranging from 350-450°F to attain a viscosity sufficiently low for pumping. Usually, gear pumps are used and the lines are jacketed. Steam is the most common heating medium, although hot-oil systems and diphenyl-vapor heating systems also are employed. The asphalt is held in storage tanks at the temperature required for the particular type of converting equipment and either pumped to the treating tanks of the machine with overflow provisions back to circulating tanks, or pumped to the machine intermittently.

Asphalt application equipment in general is similar to the equipment used in the waxing and hot melt coating industry, with the exception that frequently the construction is much heavier and built to handle much wider sheets. The simplest type consists of a submerging bath, or flooded nip, followed by a pair of squeeze rolls. Such machines are adapted only to saturating, The type most generally used consists of a pick-up roll, turning in a bath of molten asphalt, which carries the asphalt either to a transfer roll or directly to the underface of the sheet being treated. The excess is then removed and metered to the desired weight by means of a pair of squeeze rolls or by means of rods or blades. Usually, the pick-up bath, the pick-up roll itself and the metering equipment all are heated so as to maintain a suitable application temperature and asphalt viscosity. A typical asphalt laminating machine is shown in Fig. 8-1.

For laminating, one or both meeting faces of the sheets are coated with asphalt. The sheets then enter a pair or a series of combining rolls or press rolls, which are often covered with oil-resistant rubber and are hydraulic-, weight-, or spring-loaded to achieve the required combining pressure. These combining rolls, may be heated. The laminated sheet then passes over cooling drums prior to being wound into rolls. For saturation or infusion, the press section is eliminated and the sheet passes over heated drums to drive the asphaltic material into the body of the sheet.

To produce an exposed coating, the sheet travels a sufficient distance, partly settling the coat. Then it passes over brine or water-cooled drums or is transferred by means of carrying sticks or belts through cooling tunnels to chill the coat to prevent "blocking" or adherence in the finished roll. When reinforcing material is incorporated in the laminated sheets, it is applied either to the face of molten asphalt on one of the sheets travelling over the machine or is fed continuously between the two sheets just prior to the combining rolls.

Various types of specialized machines are utilized, such as equipment for producing two saturated plies by heating a lamination to the infusing point and separating the layers[12]. Numerous patents cover methods for producing two-way stretch grades, and one par-

Fig. 8-1. An asphalt laminator and creper (*Courtesy of Thilmany Pulp and Paper Company*).

ticular group of patents[13] covers a method of creping at a 45° angle to the sheet travel, with asphalt as the adhering medium so as to produce a true two-way crepe. Standard one-way crepes are produced most frequently by means of regular creping equipment. The sheet is made to adhere to the surface of a heated polished roll by means of water, glue, or other adhesive, and is removed in creped form by a doctor blade riding against the surface of the creping roll.

Other auxiliary equipment may consist of slitters or circular cutting blades used to divide and trim the web into a number of specified rolls of smaller width. In some cases, the machine is followed by formers and folding rolls to produce sheets with prefolded flanges for special uses, such as rock-wool backing. Other auxiliary equipment consists of the usual rewinders and sheeters common to the paper industry and automatic rewinders to produce measured length rolls for small-consumer use, as for the building and lumber-supply industry. Some asphalt converting includes dual operations applying wax or other material to one face or one ply and asphalt to the other.

Grades and Applications of Asphalt Papers

A complete tabulation of all the grades of paper produced by the asphalt-paper industry and their applications is almost impossible. Therefore, only typical applications of some of the more common combinations are discussed here.

Saturated and Infused Sheets

Sheets that have been saturated or infused with asphalt (short of an exposed coating) are not waterproof and water-vapor proof, and are used principally where water shedding, resistance to softening by water, and scuffing and wear resistance are desired. In certain cases, as in the "breather sheet" or sheathing paper on the outside of buildings, both the water-vapor permeability and the water shedding properties of an asphalt-saturated sheet are desirable in order to resist occasional wetting from rain but still permit the free exit of any water vapor trapped within the wall structure or insulation. A considerable part of asphalt-infused paper is used within the industry itself for recombining as laminations to enhance the wear and wetting resistance of laminated sheets. Often asphalt machines have the capability to saturate one or both plies and laminate in a single pass over the machine. The weight of asphalt in infused sheets ranges between 20 and 60% of the original weight of the base paper. A typical use is for bottoms of fiber drums where the outside surface is protected from excessive abrasion by means of a light asphalt infusion.

Laminated Sheets

Laminated combinations represent the most diversified range of weights and types of asphalt papers. So-called plain duplex sheets consist of a sheet of kraft, a layer of asphalt, and another sheet of kraft designated in order by numbers indicating basis weight. For example, 30-30-30 indicates 30 lb of kraft, 30 lb of asphalt, and 30 lb of kraft in that order. The lightest combinations do not run much below 25-25-25, and the heaviest combinations rarely exceed 90-90-90. Such sheets, if properly designed, are excellent as vapor barriers and are used wherever protection against the loss or gain of water vapor is desired or where protection against water is needed. Basis weights of paper in the combinations are selected largely according to strength requirements, whereas asphalt weights are selected largely on the basis of requirements that pertain to water-vapor resistance.

WATER VAPOR TRANSMISSION RATE
GRAMS PER 100 SQ. IN. PER 24 HOURS

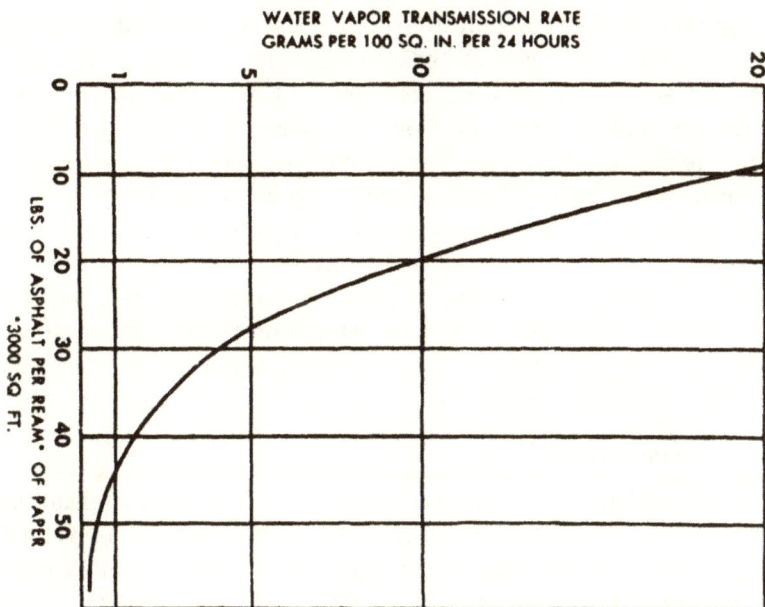

Fig. 8-2. The relationship between weight of asphalt, as a paper laminant, to water vapor transmission rate (*Courtesy of Thilmany Pulp and Paper Company*).

Figure 8-2 shows the relationship between the average water-vapor transmission rates (MVTR) (as tested in the General Foods cabinet[14]) and the basis weight of asphalt in lb/3000 ft² ream in a regular asphalt duplex combination. This curve represents average values for machine finished kraft papers. If smoother surfaced or denser papers are used, lower basis weights are needed to obtain the same resistance to water-vapor. Likewise, if rough-textured (e.g. machine-creped) sheets are used, several times the weight of asphalt may be necessary to obtain equal water-vapor resistance.

Plain duplex grades of laminated asphalt papers are used frequently as vapor barriers on the warm side of insulation in housing, in the walls of fiber drums as a vapor barrier for the packaging of moisture-sensitive materials such as photographic printing paper, and as protective overwraps on coils of tinplate, small machinery and equipment. Incorporation of aluminum foil in such laminated grades to enhance the decorative and water-vapor resistance properties is at times advantageous.

Reinforcing strands are added to asphalt duplex combinations where handling requires good tearing resistance. For reinforced asphalt papers, the basis weight of asphalt must be increased to enrobe completely the reinforcing strands and prevent wicking of moisture along and across the reinforcement itself. Common practice therefore requires utilization of three or four times the basis weight of asphalt in reinforced grades to produce moisture-vapor resistance comparable to that of plain duplex grades. Typical applications of reinforced sheets are for protection against rough handling encountered in shipping, and for tear-resistant protective wrappers for outdoor storage and open transporting of goods. Figures 8-3, 8-4, and 8-5 show several applications of shroud wrapping whereby products are wrapped and stored outdoors in open warehouse areas to await delivery in factory-fresh condition.

Asphalt Coated Sheets

Because asphalt-coated sheets are adapted to individual applications no average basis weight can be given for the paper normally used. However, comparable water-vapor transmission rates of gloss coats are obtained at basis weights $\frac{1}{3}$ to $\frac{1}{2}$ of those required for laminations because of the elimination of the possibility of cross wicking common in laminated form. As a backing and vapor barrier for rockwool insulation, coated sheets are often purchased by the insulation

Fig. 8-3. Packing and unpacking time is greatly reduced with preformed shrouds for coils of wire and other rolled steel products. (*Courtesy of Thilmany Pulp and Paper Company*).

Fig. 8-4. Because of limited storage space, engines are stored outside on pallets and covered. Cover is stapled to pallet when in place. (*Courtesy of Thilmany Pulp and Paper Company*).

manufacturer in the form of prefolded, flanged backing suitable for application to standard stud widths. Asphalt coated sheets, once commonly used as one layer in multiwall bags for protection against excess free acids in fertilizer, and to prevent moisture from penetrating the fertilizer bag where it might give rise to objectionable caking, have been replaced by polyethylene-coated papers. Asphalt-coated sheets do not replace asphalt-laminated papers because of their limitation as to appearance, blocking at high temperatures, and the possibility of rub-off, but where application permits, they provide excellent low cost protection at minimum asphalt weight.

Creped Sheets

Creped grades of asphalt papers find their widest use in specialized wrapping and packaging where a good measure of water or water-vapor resistance is desired along with properties that enable the sheet

Fig. 8-5. Substantial savings are achieved by storing and shipping grinding machines outdoors. (*Courtesy of Thilmany Pulp and Paper Company*).

to stretch and conform to the contours of irregular objects. Typical uses of creped waterproof papers include case liners, bale shrouds, coil wrapping, and interior packing to protect nursery stock and plant material. Creped grades also are used as temporary shrouds, in place of drop cloths during painting and sand blasting operations, and in dry waxed form for making disposable shower slippers and beef carcass bags.

Special Uses

Although the volume may be smaller than in the protective packaging field, asphalt papers find wide use in industrial fabrication and a few illustrations serve to show the scope of use in this field. Asphalt laminated sheets of various kinds are used in the construction of dry batteries, as waterproof casings and cell separators, and in certain types of electrical coils as separators between the primary and secondary windings. Reinforced asphalt duplex sheets form the base for special water-resistant gummed tapes and for lining materials for boxcars. Plain asphalt duplex sheets are used for waterproof tube- and can-winding stock, for cable wrap, and for waterproof tubular concrete forms.

POLYOLEFIN WATERPROOF PAPERS

Because of superior physical and chemical properties, polyethylene and to a lesser extent other olefins and polymers have supplemented or replaced asphalt as the proofing material in a considerable section of the waterproof paper industry. Asphalt is still widely used because it provides the greatest economy in the production of equivalent water and water-vapor barriers when its properties are not objectionable. An example is the building paper industry. Where low temperature flexibility, oil resistance, low odor, compliance with the Federal Food, Drug, and Cosmetic Act, or similar requirements must be met, the polymers are preferred in spite of additional cost. At times, the most economical combination consists of an asphalt lamination further protected on its surfaces with lightweight polyolefin coats.

Materials and Equipment

Polyolefins and other polymers in their low molecular weight, nonviscous form or in blends with wax, and rosin compounds or other hot melt combinations can be applied with open pan-type equipment similar to that used for the application of asphalt. Melt temperatures during application frequently must be in the range of 300 to 350°F to produce a viscosity sufficiently low for this mode of application. Barriers produced in this way usually are not ideal because of the limitations of such blends in low temperature flexibility, oil resistance, and heat bleed. The most commonly used polymers are extrusion grades of polyethylene and polypropylene, although the bulk of the production is polyethylene in the low and medium density ranges. Where appearance is important, polyethylene pigmented with titanium dioxide frequently is used. When prolonged exposure to the elements is expected, polyethylene coatings pigmented with carbon black perform well. Papers of almost any type can be utilized, whether plain, stretchable, creped, or prelaminated with other materials, and at times prereinforced. Reinforcements, when applied during the polymer coating or laminating operation, are usually glass fiber, or preformed scrim.

Electrically heated extrusion equipment is used almost universally for the application of polyolefins to paper. This equipment has been discussed and illustrated widely in the literature. All producers of polyolefin resins have available bulletins on extrusion coating. The annual issues of *Modern Plastics Encyclopedia*[16] are good sources of additional information. Resin is received in diced form, usually in bulk containers or air-unloading trucks or rail cars. The resin may be air-conveyed to the hoppers of the extrusion coating equipment. The paper web handling equipment is similar to that for other types of paper converting with few exceptions. At times, prepriming stations are supplied to apply material to the surface of the paper to promote bonding. Chill rolls at the nip following the hot extrusion application are provided with a matte finish if gloss is not desirable. Multiple unwinding sometimes must be provided, as well as suitable carrying rolls leading to the extrusion head, if the resin is to be laminated between two paper plies. Similarly, special arrangements must be planned if foil, reinforcements or similar components are to be combined with the paper and the resin at the nip following the extruder.

Combinations and Uses

All the variations of waterproof paper combinations discussed under asphalt grades, with the exception of infusion or saturation, can be produced by means of the extrusion technique. The simplest structure employing polyethylene, and perhaps the most versatile, consists of a single sheet of paper coated with polyethylene. At minimum extrusion basis weights of 4-6 lb (24 × 36—500 sheets), such single-ply coated grades are considered only water-shedding or water repellent, and possess MVTR's of approximately 3.0g/100 in.^{2}/24 hr. One-mil coatings (14.4 lb/ream) are considered water resistant and produce MVTR's in the range of 1.2 to 1.4 g.[14] Coatings up to 2 mils are common; those up to 4 mils appear in exceptional cases. Single-ply polyethylene coated papers are sold for such special end uses as battery tube stock where chemical inertness, high dielectric strength, and good moisture barrier properties are required. Polyethylene coated papers are made to a Federal Ordnance specification[15] and are used for the wrapping of oiled machine parts. Other uses of polyethylene-coated papers include liners of spirally wound fiber oil cans, ream wraps, and heat-sealable waterproof pouches.

Another grade consists of two sheets of paper laminated together with a layer of polyethylene. MVTR's comparable to coated grades are possible with polyethylene laminations when smooth densified substrates such as supercalendered papers are used. Table 8-3 gives more complete information on the protective properties of polyethylene coated and laminated grades. Polyethylene-laminated papers are used for automobile door panels and for liner stock for fiber drums because of their permanence and stability over a broad temperature range. Such laminations can be reinforced on the extruder; however, prewoven scrim is invariably used because of the proximity of the extruder head to the laminating nip, which prevents the use of weaving equipment.

Reinforced polyethylene-laminated papers are used for carpet wrap, magazine and book wrap, brick wrap, and wherever high tearing resistance is required in addition to the usual properties of plain polyethylene grades. Any number of combinations can be made by incorporating polyethylene, reinforcement, and paper. In addition, reinforced polyethylene-laminated papers are overcoated on one or both sides for the wrapping of coils of oiled steel and prefabricated structural wood beams. The inner coating serves to prevent steel slushing oils and wood beam sealers from staining the wrapper, thus reducing the effectiveness of the treatment and spoiling the appearance of the wrapper.

With the advent of revolutionary handling methods, such as unitized packaging, increasing quantities of a wide variety of merchandise are transported in "open top" gondola or flat car shipments, or are

TABLE 8-3

Properties of Polyethylene Coated and Laminated Kraft Papers

(*Courtesy Thilmany Pulp & Paper Co.*)

Poly weight, lb	Water resistance, hr[1]	MVTR, g^2	DTE oil 30 SAE (lube oil)	$O_2{}^3$	$CO_2{}^3$
Coated					
4	10	3.0	15 hr	No resistance	No resistance
8	20	2.0	24 "	374 cm³	1870 cm³
10	24	1.5	50 "	302 "	1510 "
15	36	1.0	100 "	230 "	1150 "
20	72	0.81	150+ "	180 "	900 "
25	100	0.72	2 wk	130 "	650 "
30	120+	0.65	2+ "	109 "	547 "
45	120+	0.45	2+ "	60	290 "
Laminated					
7.5	5	4.0	10 hr	No resistance	No resistance
10	15	2.0	24 "	1150 cc.	5600 cc.
15	24	1.2	36 "	448 "	2240 "
20	48+	0.8	48+ "	200 "	1000 "

The above figures are for polyethylene of 0.92 density. All above resistances increase with increased polyethylene density. Results can vary appreciably depending upon the roughness of the substrate.

1 TAPPI Procedure T-433 m-44 Dry Indicator Method.
2 MVTR by General Foods Method. Reported in grams transmission per 100 in.²/24 hr at 100°F.
3 Gas resistance results obtained on 100 in.²/24 hr (similar to MVTR).

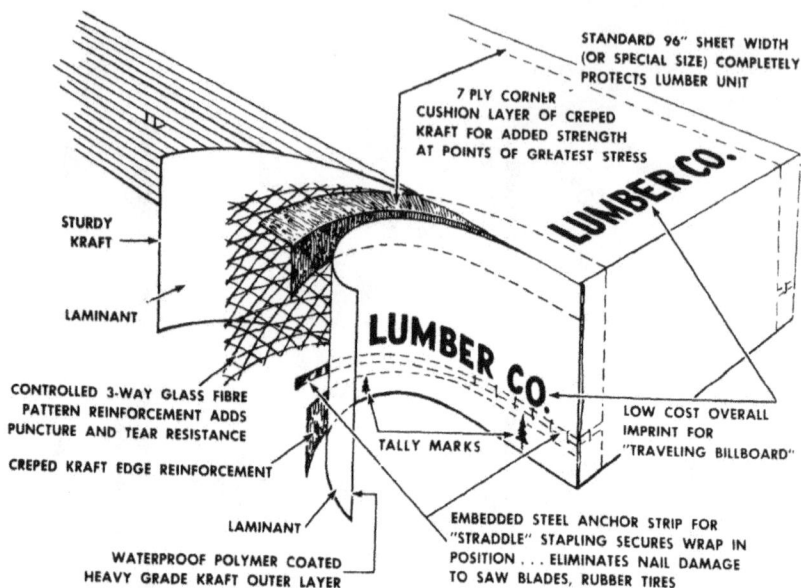

STANDARD 96" SHEET WIDTH
(OR SPECIAL SIZE) COMPLETELY
PROTECTS LUMBER UNIT

7 PLY CORNER
CUSHION LAYER OF CREPED
KRAFT FOR ADDED STRENGTH
AT POINTS OF GREATEST STRESS

STURDY
KRAFT

LAMINANT

CONTROLLED 3-WAY GLASS FIBRE
PATTERN REINFORCEMENT ADDS
PUNCTURE AND TEAR RESISTANCE

CREPED KRAFT EDGE REINFORCEMENT

LOW COST OVERALL
IMPRINT FOR
"TRAVELING BILLBOARD"

TALLY MARKS

LAMINANT

WATERPROOF POLYMER COATED
HEAVY GRADE KRAFT OUTER LAYER

EMBEDDED STEEL ANCHOR STRIP FOR
"STRADDLE" STAPLING SECURES WRAP IN
POSITION . . . ELIMINATES NAIL DAMAGE
TO SAW BLADES, RUBBER TIRES

Fig. 8-6.   The internal structure of a lumber wrapper (*Courtesy of Thilmany Pulp and Paper Company*).

Fig. 8-7.   The laminated wooden beam industry uses a polyolefin coated, reinforced, nonasphaltic protective wrapper for the wrapping of beams in storage and in transit (*Courtesy of Thilmany Pulp and Paper Company*).

stored outdoors in temporary "warehouses." Specially designed reinforced, polyethylene-coated, waterproof wrappers have been developed for open air, long-term storage as used in the wrapping of unitized dimensional lumber and skids. Polyethylene-coated asphaltic laminations combine the low-cost moisture protection of asphalt with the nonstaining properties and attractiveness of polyethylene. Such wrappers provide improved water shedding, excellent moisture barrier, good outdoor weatherability, and readily printable surfaces, which become customer travelling billboards. Figure 8-6 shows the internal structure of a lumber wrapper with special areas of additional reinforcement and the inclusion of a metal band for fastening down the wrapper. Figure 8-7 shows a wrapped prefabricated wood beam on a flat car.

Many standard asphalt-laminated grades mentioned in this chapter frequently are modified by replacing the normal asphalt weight with a much lower weight of microcrystalline wax to produce lightweight wrappers of comparable MVTR.

REFERENCES

1. Federal Spec. for Paper; Kraft, Wrapping, Waterproofed, UU-P-271d.
2. U.S. Army, Spec. No. 100-14-A, General Spec. for Packaging and Packing Overseas Shipments (Issue of Feb. 15, 1943).
3. Herbert Abraham. *Asphalts and Allied Substances*, 5th Edition, Vol. I, D. Van Nostrand Co., Inc., New York (1945), p. 49.
4. T. R. Williams, British Patent 10,774 (1845).
5. W. H. Childs, U.S. Patent 361,050 (1887).
6. J. Arkell, U.S. Patent 670,393 (1901).
7. A. Wendler, German Patent 222,959 (1909).
8. *Communication* from The Waterproof Paper Manufacturers Association, New York.
9. American Society for Testing Materials Standard Method D-36-26, Volume IV (1955), p. 1030.
10. American Society for Testing Materials Standard Method D5-52, Volume IV (1955), p. 1003.
11. "Signode Patent Grain Door," Brochure of Signode Corporation, Chicago, Ill.
12. W. M. Wheildon, U.S. Patent 1,595,637 (1926).
13. W. C. Kemp, U.S. Patents 2,008,181 (1935); 2,008,182 (1935); and 2,071,347 (1937).
14. M. Yesek, *Packaging Parade* **24**, No. 8:103 (Aug. 1956).
15. Military Specification MIL-B-121B, dated 21 May 1959.
16. *Modern Plastics Encyclopedia* (annual), McGraw-Hill, New York.

chapter 9

Waxed Papers

A. M. WORTHINGTON

The development of waxed paper, which dates back to 1890 may be ascribed to two important industrial trends: the manufacture of paper from wood pulp and the manufacture of paraffin wax from petroleum. These two innovations, which occurred at about the same time, made possible the production of paper designed especially to be waterproof and water-vapor proof. Waxed papers were introduced commercially by the National Biscuit Company for the inner wrap of their Uneeda Biscuit package; they were used for the inner wrap for Crackerjack boxes at an early date.

Paper, which is a mat of felted cellulose fibers of higher or lower degree of purity, embodies the properties of the individual fibers themselves, one of which is that of being hygroscopic, or being able to absorb moisture from the air. Cellulose also permits desorption of the moisture it holds to the air, if the air is dry. It is a carbohydrate, containing hydroxyl radicals that are hydrophilic. When paper is exposed to moist air, these OH groups attract water molecules; when paper is exposed to dry air, the water molecules are again given off to the atmosphere. The amount of water that can be absorbed is 6 to 8%; water is absorbed more readily than it is given up.

Natural waxes had been used for many centuries as decorative or protective coatings. That waxes could be used to impart waterproofness was commonly known. When paraffin wax was developed, its cheapness and availability made it ideal for impregnating and coating paper to render it waterproof and water-vapor proof. Waxed paper and paperboard are now widely used for wrapping and packaging materials where protection from moisture or dryness is desired. Perhaps the largest and most important group of materials thus protected are foods—bread, cakes, meat, sandwiches, delicatessen foods, butter, cheeses, frozen foods, dehydrated foods, crackers, tea, coffee and candy.

MANUFACTURE

Most of the paper used for the manufacture of waxed paper is made from bleached sulfite pulp. To a lesser extent, unbleached sulfite and sometimes sulfate pulp are used. The sulfite sheet may be either plain or loaded with an opaque filler. Greaseproof krafts, glassines, manilas and several grades of twisting papers can also be used.

Where a grade of paper such as glassine, which is practically impervious to wax, is converted, the process is simply one of coating either one or both sides of the sheet with the wax forming a surface film. The waxing of sulfite and similar grades of paper is a process of either wax penetration or wax penetration and wax coating. The former is known as dry waxing and the latter as wet waxing. Figure 9-1 illustrates the differences. In dry waxing, the wax is driven into the sheet

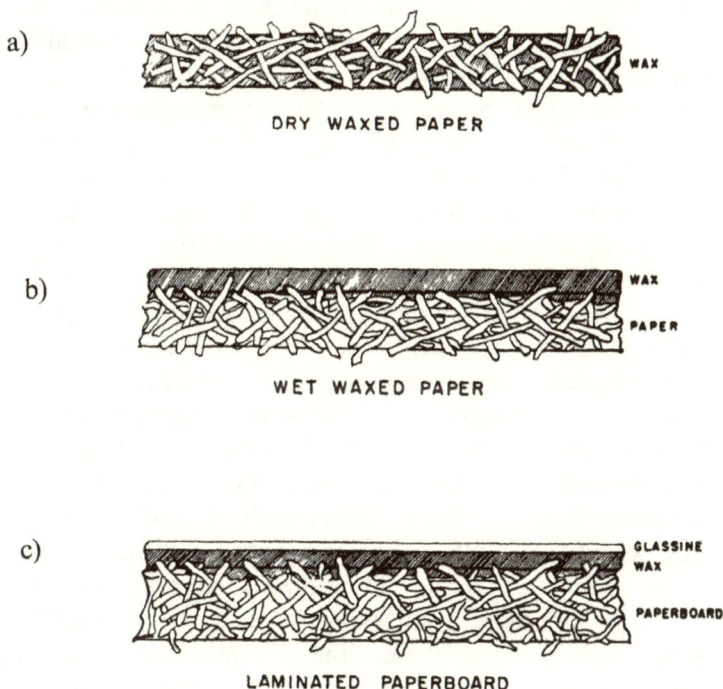

a)

DRY WAXED PAPER

b)

WET WAXED PAPER

c)

LAMINATED PAPERBOARD

Fig. 9-1. Three methods of waxing, (a) Dry waxed paper; (b) wet waxed paper; (c) laminated paperboard (*Courtesy of Modern packaging Magazine*).

and fills the spaces between the fibers, with little wax being left on the surface.

In the case of wet waxing, most of the wax remains on the surface as a continuous film, with just enough penetrating into the sheet to bond the film. Water passes through paper by way of the spaces between the fibers, whereas water vapor is transmitted both through these spaces and through or along the fibers themselves. If the surface of a sheet of paper is not completely covered with wax, some of the fibers protrude and act as wicks for the transmission of water vapor through the sheet. Thus, a dry waxed sheet is waterproof, and a wet waxed sheet both waterproof and water-vapor proof. Lamination, as shown in the third part of Fig. 10-1(c) provides water-vapor-proofing with no surface wax film, by placing a film of wax between two similar or dissimilar sheets that acts as a bond between them. Waxed papers now commercially produced have a water-vapor transmission rate as low as 0.2 g/100 in.^{2}/24 hr. at a temperature of 100°F and a relative humidity of 95% on the wet side of the sheet, or 3.1/m^2/24 hr.

Most waxed paper is made on waxing machines. The history of such machines goes back to 1866, to Stuart Gwynn of New York, who was the first to patent the use of paraffin for impregnating paper. In 1878, Siegried Hammerschlag, who has been called the father of the coating section of the waxing machine, applied a wax coating to one side of a sheet by running it over the top surface of a steam-heated roll revolving partly submerged in a bath of wax. The speed of the roll was variable, which in turn varied the amount of the coating. The surplus wax was removed by scrapers. A fan cooled the waxed paper.

In 1879, Hammerschlag went a step farther and developed the principle of the squeeze roll section as we know it today. This is a very important part of the machine, as this section mainly controls the amount of wax applied to the sheet. Hammerschlag patented a machine with two vertically mounted rolls, the bottom one of hollow iron to be heated by steam and the top being made of wood or rubber. The journal boxes were in slides and were provided with weights or levers to vary the pressure of the top roll on the bottom roll. The sheet was first coated on both sides by immersion in wax. It was passed through the squeeze rolls, and then the coating was smoothed by drawing the sheet over hot pipes. A cooling fan was also used. Hammerschlag was the originator of many of the steps in wet-waxing as we know them.

The next important development occurred in 1901 when Norris and Vavra patented the use of water as a cooling medium to set the wax film. They found that the sudden cooling of the wax and paper arrested wax absorption and gave a smooth, glossy appearance to the waxed sheet. In their work, they found the most suitable wax temperature to be 150 to 175°F and the temperature of the bath of cooling water to be 32 to 40°F.

In 1915, Carter patented a machine that ran paper through a wax bath, over a set of hollow, heated, smoothing or ironing rolls (two of which were turning in the opposite direction to the travel of the sheet), and through a bath of cold water. The residual water was removed by an air blast in the direction opposite to the paper travel. In 1921, the speed of the water-cooling machine was increased by the patented device of Decker and van Sluys of vibrating rods bearing against the paper across the web, followed by a set of scrapers against the paper, and finally suction pipes between which the paper passed. All these were designed to provide more rapid water removal. In 1922, Carter patented and used a set of four cooling rolls to chill the waxed sheet. Liquid from a refrigerating plant was used to keep the rolls at the desired temperature. Further patents have not changed the basic principles of the waxing machine. It is interesting to note, however, that in connection with developments leading to increased production of waxed paper per machine, in 1928 Nunez patented a duplex waxing machine, comprising supports for a pair of dry paper rolls, a wax pan, guide rolls to direct two webs of paper independently through the wax bath, independent chilling rolls for each web, and two separate winders for the waxed paper rolls. In 1959 Yezek and Utschig[1] patented a device for putting a glossy wax coating on paper.

An exception to the usual type of waxing machine is one that can be considered almost completely as a coating machine. Either one or both sides of a web of paper can be wax-coated by using this equipment. For one-side coating, the paper passes over an equalizer rod, a highly polished and accurately ground steel rod, $3/16$ in. in diameter, wound with fine-diameter steel wire. This rod is driven and is fed by a driven roll revolving in a wax tank. A set of rods comes with each machine wound with wire thicknesses ranging from 0.003 to 0.050 in. The amount of coating is regulated by the wire gage, the finer the wire the lighter the coating. For coating both sides of the sheet, the paper passes over one equalizer and is then reversed in direction and passed over a second equalizer, operating from a second wax bath. The sheet is then drawn over a cold roll to chill and solidify the wax coating.

Another development is the use of petroleum products other than paraffin wax for the impregnation of paper to provide waterproofing and grease resistance. These products are petrolatum, sometimes called soft-type microcrystalline wax; white oil; and pale oil. Paper treated with these waterproofing agents is comparable to dry waxed paper only, but the distinctive properties of these substances and their ease of application have caused them to replace waxed papers for some purposes, and to extend the use of petroleum-product-treated papers to new fields.

Base Stocks Required

The principal advantage of a sulfite sheet for waxing is that it can be tailor-made to fit the specifications laid down for the uses for which it is intended. Several properties are controlled to produce a sheet that will respond in the ways desired when the paper is run through the waxing machines.

The main requirements of the raw or base stock for a sheet of sulfite or sulfate waxing paper are:

1. Proper finish.
2. Correct texture or hardness of the sheet.
3. Adequate physical strength of the sheet.
4. Proper density or ratio of basis weight to thickness.
5. Proper moisture content.

Finish. The finish of the raw stock may be low, medium, high dry, and high wet, for machine-finished sheets. A high dry finish is made by passing the sheet through all the nips of one or two calender stacks at the paper machine, with no steam shower on the first stack. A high wet finish paper is made by using a steam shower on the first stack to put moisture into the sheet and thus obtain a higher finish. In addition, the sheet may be supercalendered for an even higher finish.

Of these finishes, the first three are part of the requirements of a sheet for dry waxing. The lower the finish the greater the ease with which the wax penetrates into the sheet. Decreasing wax loads are obtained as paper with a higher degree of finish is put through the waxer. The two other finishes—high wet and supercalendered—are preferred where the sheet is to be wet or full waxed. Such finishes resist penetration of the wax, provide a smooth base for the external wax film, and thus help to impart a high gloss to the waxed sheet.

Texture. Closely related to finish in influencing the behavior of the wax during the waxing process is the hardness or texture of the sheet. Everything in the papermaking process affects texture, from the raw pulp to final calendering. Strong bleached, unbleached sulfite pulp, or a mixture of the two are generally used in the beater furnish, depending on the color and brightness desired in the finished sheet. In some cases, sulfate pulp, or a portion of it, is used. The sheet must be well formed, but the fibers should not be too short, as the sheet is to be used for wrapping purposes and must not be too brittle. One way of judging the hardness of a sheet is by the bursting strength-tear relationship. Low burst with high tear indicates a soft sheet, i.e. one with a tendency to absorb too much wax. Such a sheet, when waxed, would be reduced in strength as the large amount of wax between the fibers would have a lubricating effect that would destroy the bond between them. High burst with very low tear indicates a brittle sheet, i.e. one that would tend to break and tear easily when used for wrapping.

Strength. A sheet with the proper strength and degree of hardness for correct waxing would be (a) for dry waxing—one with a bursting test of 50% of the basis weight in numerical value, and an average tear test (averaging the tests with and across the grain) equal in grams to the figure for the basis weight of the sheet (basis weight: ream weight of 24×36 in.—480 sheets); (b) for wet waxing—the sheet would need to be harder and the bursting strength requirement would be 60% of the basis weight with the tear test almost the same as before. The porosity of the sheet was formerly thought to be a factor but more recent work has not shown any direct influence of this property on the waxing of the sheet.

Soft sheets are indicative of insufficient treatment in the beaters. Such sheets may be greatly improved for waxing purposes by supercalendering for giving them a high finish to reduce wax penetration. In connection with the question of soft texture, it should be noted whether the sheet under consideration contains opaque filler. Such a sheet would have a softer texture than one made in the same way without the filler, and allowances would have to be made accordingly in the hardness requirements. The filler is mineral, and since it occupies spaces between the fibers, it tends to reduce the penetration of the wax.

Density. The density of the sheet, which is the ratio of the basis weight to the thickness (caliper), is also important. Density is related to the other properties of the sheet already described, but it can also serve as a guide as to the waxing qualities of the sheet because the higher the density, the more the sheet will resist wax penetration.

Moisture. The sheet should carry as much moisture as is practical.

For dry waxing, where wax penetration is desired, 6 to 8%; for wet waxing, where penetration is to be restricted, more moisture, namely 7 to 9%, is required.

PRINTING OF WAXING PAPER

In many of the uses of waxed paper it is advantageous or necessary to have the paper printed, as a descriptive and advertising device for the product that is to be wrapped in the paper. Where printing is to be done, this step is carried out before the paper is waxed. Because of the texture of the sheet of waxing paper and the fact that it is subsequently to be waxed, all types of inks are not suitable. The inks and the printing methods described in the following paragraphs are among those generally used.

Oxidizing Inks. Oxidizing inks may be used in any of the three types of printing: letterpress, lithography or offset, and intaglio or rotogravure. They harden because the vehicle is a drying oil or drying oil-modified resin, which oxidizes in the air, aided by catalysts that are usually organic compounds of cobalt, lead, and manganese. These inks take 1 or 2 days to harden suffiiciently so that they do not bleed when the paper is put through the waxing machine.

Immediately after being printed, the back of the sheet may be lightly waxed before it is rewound, to prevent offset. This is usually done by applying the wax by means of a roll, and then chilling the sheet by passing it over a cold roll, corresponding to a one-side coating method. The wax used is of the same melting point as that to be used later when the sheet is waxed.

Aniline Inks. Aniline inks and presses were developed so that paper could be printed and converted to other forms in one continuous operation. This type of printing is now widely used for bread wraps. Aniline printing plates have an extremely long life: as many as a million impressions can be made from a single rubber plate. The solvents necessary with these inks are often toxic and inflammable and must be vented from the press room for the sake of health and safety. Binders of various types are used with the solvents.

Heat-Drying Inks. Heat drying inks were brought out in answer to a demand for higher production speeds for fast quality letterpress printing on web-fed presses. They use a binding resin in a solvent such as a mineral oil fraction that is essentially nonvolatile at room temperatures but very volatile at high temperatures. Because of these properties, the inks dry almost instantly upon the application of heat, which can be from an open flame. This literally burns away the solvent in the ink, leaving the binding resin and pigment on the paper. The flame temperature may be in excess of 1000°F. The inks may be dried by passing the web over large steam-heated drums, or through hot-air chambers with a large amount of superheated, unsaturated air. Combinations of these methods can also be used. The paper then passes over chilling rolls to bring the temperature back to normal. These inks are being used in bag printing.

Moisture-Set Inks. There are a comparatively new development. The principle of drying involves neither oxidation nor evaporation of solvents. It is based on the principle of precipitation. The ink consists of a pigment or pigments dispersed in a vehicle made of synthetic resin dissolved in a high-boiling solvent, the resin being soluble in the solvent and a limited quantity of water. The resin is insoluble in a greater quantity of water mixed with the solvent and is precipitated in the presence of this added water, thus "drying" the ink film. Water may be added in one of four principal ways: (a) spraying after printing, (b) adding to paper before printing, (c) drawing from the moisture in the air, or (d) drawing from the moisture already in the material. These inks are being used on chewing gum wraps, bread wraps, and for similar applications where an odorless ink is necessary. Paper thus printed can be waxed within 2 hours after printing.

WAXES

Paraffin Wax

Paraffin wax, most commonly used in the waxing of paper, is mainly derived from high-boiling fractions of petroleum. Each year nearly one billion pounds of this wax are produced in the United States, of which 80% is used for paper, 10% for candles, and 10% for other purposes. It is not a chemically pure compound, but a white, translucent, tasteless, odorless, solid, that consists of a mixture of solid hydrocarbons, chiefly of the methane series, when in a refined state. The crude grades are odorous, greasy, and contain volatile constituents, and as a result are not important in connection with waxed papers. Refined paraffin wax is soluble in benzol, ether, chloroform, carbon disulfide, carbon tetrachloride, turpentine, petroleum, and fixed oils. It is insoluble in water and cold alcohol.

Microcrystalline Wax

Microcrystalline waxes, sometimes incorrectly called "amorphous waxes," occupy an important place in the waxing of paper. They are paraffinic, crystalline waxes, but their crystals are much smaller than those of ordinary paraffin wax. These waxes can be used alone or blended with paraffin waxes. Microcrystalline waxes are made up of the methane series hydrocarbons of high melting points and high molecular weights. The work of Ferris, Cowles, and Henderson and also that of Buchler and Graves shows that the crystals are in needle and malcrystalline forms. The formulas range from $C_{34}H_{70}$ to $C_{43}H_{88}$. The melting points may run from 140 to 200°F. They are not white but of varying degrees of brownish-yellow in color. Compared to paraffin they are tougher, have higher ductility and tensile strength, have greater tackiness, and are less lustrous and greasy. They have higher viscosities in the molten condition.

Paraffin vs. Microcrystalline Wax

These two waxes are widely used in the paper industry for the water-proofing and water-vapor-proofing of paper. They are economical, easy to handle, and satisfactory in use. Each type has its advantages and disadvantages compared to the other. Blends of the two types may be made to produce a combination of desirable properties of both, or either may be used as a base to which other ingredients may be added to modify the original properties.

The advantages of paraffin wax are whiteness, producing practically colorless wax films; hardness; non-tackiness; ability to achieve high gloss in a film; and low viscosities when molten, for production of lower wax loads. The disadvantages are brittleness, poor aging qualities at elevated temperatures, and low grease resistance.

The advantages of microcrystalline wax are greater ductility over a wide temperature range, making the wax useful where waxed sheets may be crumpled or creased, or used in the wrapping of quick-frozen foods; tackiness, making this wax useful in the laminating field; markedly greater grease resistance; higher viscosities in the molten state for high wax loads; good aging qualities at elevated temperatures; and good stability to ward oxidation. The main disadvantage of this wax is its color, restricting its use to papers where light appearance is not so important, for coating dark-colored sheets, or for use in inner wraps.

Waxes do not in themselves render a sheet of paper greaseproof;

greaseproofing is generally the function of the base sheet of paper. Microcrystalline waxes do have the property of imparting much greater grease resistance to paper than does paraffin, and a heavy coating of wax enables a sheet to pass some specifications for greaseproofness.

OILS

Petrolatum

Oils as additives have a softening effect on paper. Petrolatum, or soft type microcrystalline wax, is a purified semisolid mixture of hydrocarbons, of salve-like consistency, and transparent in thin layers. It is manufactured in various grades, ranging in color from dark-green to white. The highly purified grades are free from odor and taste, and are suitable for use in food wraps because they do not become rancid. Petrolatum is freely miscible with other petroleum products such as waxes and oils, and these are often used to modify the characteristics of petrolatum as required for various treated papers. Applied to paper, it has a softening effect. Compounded with waxes and added to paper, a harder drier sheet is produced.

Petrolatum is a colloidal system of two phases, the internal phase being liquid hydrocarbons and the external phase solid hydrocarbons. The melting point ranges from 110 to 137°F by the ASTM method. Its stability to oxidation is comparable to that of a fully refined paraffin wax.

White Oils

These are paraffin oils that have been additionally refined to remove practically all reactive and unstable components. They are light-bodied, extremely light in color, and have high stability against discoloration and rancidity.

Pale Oils

These are also paraffin oils. They are of low viscosity, and are practically free from odor. They may be used where the degree of purity need not be as high as with applications requiring white oil.

WAX FORMULATION AND COMPOUNDING

The waxing of paper is usually carried out with a paraffin or microcrystalline wax of the desired melting point with no other added substances. Where certain effects not obtainable under such conditions

are desired, compatible compounds may be added to the wax in the
molten condition. Any one or more of the properties of petroleum
waxes can be modified by the use of additives, and marked changes
can be obtained by employing blends of the waxes with other substances. *Ductility*, *flexibility*, and *pliability* are related properties,
and can be described as the ability of the wax to be distorted without
breaking. In the case of paraffin, resistance to breaking can be increased by the addition of microcrystalline wax or petrolatum. Microcrystalline wax can be made even more ductile by incorporating an oil
or petrolatum. Polyisobutylene, a synthetic thermoplastic elastomer,
increases flexibility when added to wax formulations. It blends with
wax with some difficulty; a concentrated mixture as supplied by the
manufacturers can then be diluted in the wax bath to a content of
0.5 to 2% polyisobutylene. Cyclized rubber, added to wax, improves
pliability. Hydrogenated oils or fats, added to the wax, produce waxed paper of greater folding endurance.

Viscosity or Body

Viscosity can be increased by adding metallic soaps, such as aluminum stearate, rubber polybutenes, or polyisobutylenes, methacrylate
polymers, cellulose ethers, Acrawax C, Strobawax, or butyl rubber.
(Many of these are supplied as concentrated mixtures with paraffin,
to be diluted during the waxing process to 2 to 5% butyl rubber in
paraffin.) Increasing the viscosity of the wax decreases its penetration
during waxing.

Viscosity can be decreased by blending the wax with substances
of lower viscosity or solvents. Microcrystalline wax can be reduced
by the addition of a small amount of paraffin. Penetration is increased by applying the wax at a higher temperature and thus lowering the
viscosity. The addition of small amounts of fatty material such as
stearic acid (5% or less) improves penetration.

Adhesiveness

This property may be increased where it is desired to have a tackier
wax for a stronger heat seal, or a stronger bond in a laminated sheet.
For this purpose, rosin, polymerized rosin, or hydrogenated rosin are
most commonly used. Polymerized rosin is superior to the hydrogenated type. Other satisfactory materials are polyisobutylene, butyl
rubber, and cyclized rubber.

Slipperiness

This is increased by hardening the wax with some additive. Paraffin wax can be added to microcrystalline wax. In the case of paraffin, carnauba wax, or some other vegetable wax, or hard resins can be used. Albacer, a synthetic wax, added to paraffin in a proportion of 5 to 10% prevents adhesion of the wax to sticky candies.

Tensile Strength

This property of paraffin is increased by blending it with microcrystalline wax. Stearic acid and ozokerite (an amorphous natural hydrocarbon wax) can also be used.

Hardness

In the case of paraffin, hardness can be increased by the use of small amounts of the following additives: carnauba wax, Acrawax C, IG Wax S, IG Wax Z (the last three being synthetic waxes). Hardness is desired to reduce wax penetration into the sheet, and to stiffen the waxed sheet.

Melting Point

With paraffin, the melting point can be increased by the addition of carnauba wax or ozokerite.

Luster and Gloss

Carnauba wax and rosin or its derivatives added to paraffin increase luster and gloss.

Opacity

Opaque waxes may be prepared by blowing air or any other suitable gaseous medium into the molten wax to which a froth stabilizer, such as a soap, has been added previously. The minute bubbles trapped in the wax upon solidification render it more opaque. The froth stabilizer permits reheating of the wax above its melting point with retention of the greater part of the bubbles when the wax resolidifies. Other methods of increasing opacity are (1) incorporation of a suspension of titanium dioxide and a small amount of stearic acid or some substance with an acid reaction in the paraffin wax, and (2) addition of 0.5 to 5% of a hydrogenated vegetable oil. The methods of rendering waxes more opaque are patented, as are the uses of some of the

TABLE 9-1
Wax Blends

Formula No.	1	2	3	4	5	6	7	8	9	10
Paraffin	85	65	75	40	60	55	30–60	5–50	20	100
Carnauba	10			40	20		5–15			
Rubber	5	5	5	5	5	4			2–3	1–3
Candelilla		30				25				
Gum dammar			20		15					
Ester gum				15				5–75	80	
Hydrogenated castor oil						16				
Pitch							10–25			
Rosin							10–30			
Petrolatum wax								25–90		
Titanium dioxide										5–15
Sodium benzoate										5–15

substances mentioned in preceding paragraphs of this section.

Wax Blends

Examples of a few of the wax blends that have been disclosed in the patent literature are given in Table 10-1

Resin Wax Blends

Resins compatible with petroleum waxes are largely composed of hydrocarbons, such as polymerized terpenes, rubber-like polymers, and certain vinyl chloride ⁻ ins. Some normal phenol-modified resins, ester gum and fᵛ   ᴄᵤᵍo have also been ₁ ₁d to be compatible with paraffin and ..ᴇr waxes.   Under continuous heating, however, there is separation because of the oxidation of portions of the resins.   If these portions are reduced to a minimum during the manufacture of the resin, greater compatibility is achieved, and such resins are more suitable for blending with wax for coating.

The addition of 20 to 25% resin to paraffin wax increases the luster, grease resistance, and water resistance of the wax.   Such addition also reduces the tendency of the wax to smear and collect dirt, but it gives little or no improvement in the heat sealing qualities.

Ethyl Cellulose

This resin can be used together with paraffin provided a mutual solvent is also present.   Ethyl cellulose increases the toughness of wax films.   Two of the recommended formulas are given below.

Formula 1		Formula 2	
	Parts		%
Paraffin	10	Ethyl cellulose	10
Montan wax (derived		Stearic acid	45
from lignite)	10		
Ethyl cellulose	2	Paraffin	45

WAXING PROCESSES (IMPREGNATING AND COATING ON WAXING MACHINES)

The three methods of waxing by means of waxing machines are dry waxing, wet waxing, and laminating.

Dry Waxing

This method is designed to impregnate the sheet without leaving a surface film. On some machines, the sheet is unwound from the raw paper roll, and fed through the nip between two squeeze rolls, the bottom one of which is partially submerged in the molten wax. The pressure at the nip forces the wax into the sheet. A heated roll immediately following the squeeze rolls aids in the penetration. The sheet can then be passed over one or more chill rolls to cool and set the wax, or it can be passed over additional heated rolls and wound up hot, depending on the character of the impregnation desired. The finishing rolls can be used interchangeably as chill or hot rolls by having both steam and cold water connections. The first method leaves a waxy feel to the surface of the sheet; the second method gives the sheet a dry feel, as the wax is allowed an extended time for thorough penetration. To keep the wax load down on this type of machine, a three-roll squeeze section can be used. The paper passes through the upper nip, where the amount of wax transferred from the bath is less than at the lower nip.

On another type of machine, the wax is applied by an "advance roll," a heated single roll of variable speed revolving in the wax bath. The paper passes over this roll. The amount of wax applied can be increased by increasing the speed of the advance roll. At all times the peripheral speed of this roll is less than the speed of the paper, so that the wax is applied by a wiping process. The distribution of the wax can be varied by changing the arc of contact between the paper and the advance roll by raising or lowering two rolls under which the paper passes before and after the advance roll. The paper then goes through the nip of the squeeze rolls, the bottom one of which is heated to increase penetration, but is out of contact with the wax bath. For heavier impregnation, the lower squeeze roll can be partially submerged in the wax bath.

Wet Waxing

The purpose of wet waxing is to provide a continuous surface film or coating of wax either on one or both sides of the sheet of paper.

For one-side coating the method is like that for dry waxing. The wax is applied either by means of an advance roll in the wiping process without squeeze rolls, or by means of squeeze rolls without an advance roll. The sheet then passes around the chill rolls to stop penetration and keep a film on the surface. The squeeze-roll method is not applicable to tissue sheets for coating one side as the wax penetrates through the sheet too readily. For heavier-weight sheets the paper itself prevents contact of the wax with the upper squeeze roll, and a one-side coating is provided by the wax supplied by the lower squeeze

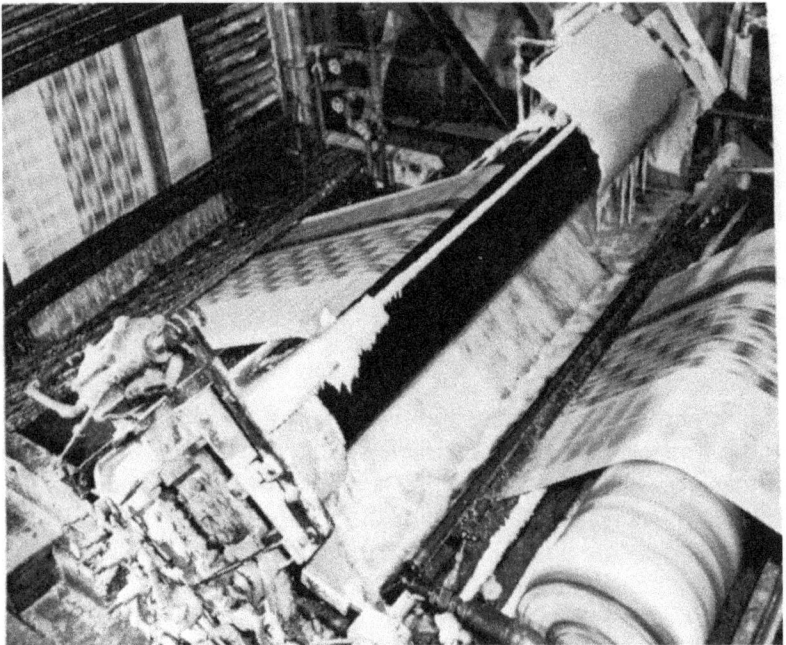

Fig. 9-2. Wet waxing operation on bread wrappers (*Courtesy of The Marathon Corp.*).

roll, which is partly submerged in the wax bath.

For coating both sides of the paper, the sheet passes under a dip roll and is immersed in the wax (Fig. 9-2). The sheet then goes through the nip of the squeeze rolls which regulate the amount of wax put on the sheet. If a heavier wax load than can be applied by the use of uncovered squeeze rolls is required, the rolls can be covered with woolen blankets. The bottom blanket is kept saturated with wax as the bottom roll revolves in the bath. The blankets cushion the sheet as it goes through, so that more of the wax is retained on the sheet, and at the same time the squeeze rolls serve to keep the wax load uniform. The sheet then passes over a hot roll to maintain the proper temperature for the polishing of the film. This is done by passing the sheet down between brass rolls revolving in the direction opposite to the travel of the paper.

The wax film is then set in one of two ways. The first and earlier method is to pass the paper around two or more cold rolls through which cold water or brine is circulated. Machines for this method are known as "cold roll machines." The second method is to immerse the paper in a bath of cold water to chill the sheet and set the wax film, followed by removing this water before winding the sheet into the finished roll. Machines for this method are known as "water waxers."

Water is removed by squeezing, blowing, suction, or scraping. In the squeezing method, the paper passes between two rubber-covered rolls like a wringer. Here the water removal is not 100%. In the suction method, perforated tubes connected to a vacuum pump are used. The sheet passes over these and the water is drawn off. The disadvantage of this method is that it is necessary to deckle the tubes in for various sheet widths. Constant checking against leaks is also required.

Two examples of the blowing methods are those patented in 1928 by Hayward and in 1937 by Potdevin. In Hayward's method, two vertical sets of box-like blowers are arranged in series on alternately opposite sides of the web of paper. The tension of the sheet prevents it from blowing away and the air blast keeps it from coming in contact with or rubbing against the blower heads, while the water is blown off the sheet. In Potdevin's method, the sheet moves between air streams directed against first one side of the sheet and then the other to vibrate the sheet, and remove the water.

The scraping method illustrated is that patented by Howard in 1939, in which the Howard water-removal tower is used. After the sheet has been submerged in the water bath, it goes upward past two water-removal doctors. These remove the major part of the water, leaving only enough to act as a cushion. The sheet goes over a roll at the top of the tower, then downward past two banks of knives, twelve to the bank. The top bank removes the water from the lower side of the sheet, and the bottom bank does the same for the upper side. The water still being carried by the web protects it from too close contact with the knives. These blades are set at an angle and have edges so tooled that drainage is effected after the water is removed. From the tower the sheet may go to cooling rolls and then be rewound, or it may be rewound directly. This apparatus efficiently removes water from the sheet at speeds of over 1000 ft/min.

Carton Waxing

Carton waxers are like water waxers. The cartons are printed and stamped out before waxing, although sometimes they are waxed in sheet form. They are fed in, one at a time, by the machine operator. After the water bath the cartons pass through a series of felt squeeze rolls to remove the water. Endless belts conduct the cartons through the waxing machine.

Wax Load

Dry-waxed sheets usually have a wax load of 15 to 20% of the dry sheet. With wet-waxed sheets, the load is 40 to 50% or even higher, based on the dry weight of the sheet. In a dry-waxed sheet, it is desired to have the wax impregnated into the sheet. When wet waxing, only enough wax should be impregnated to anchor the surface film to the sheet.

FACTORS CONTROLLING AMOUNT AND TYPE OF COATING

Speed of Sheet

The speed of the sheet should be equal to the peripheral speed of the squeeze rolls at all times. With wet-waxed sheets, having a high percentage of wax on the surface, the wax film should be carried through the nip. Since hot wax acts as a lubricant, the paper has a tendency to pass through the nip of the squeeze rolls either too fast or too slow, when proper attention is not given to the running of the waxing machine—too slow if too much friction is applied to the dry roll shaft compared to the power of the winder, and too fast if too little friction is applied.

Speed of Machine

At higher speeds, the sheet carries more wax to the nip of the squeeze rolls, and at such speeds there is also less time for penetration of the wax into the sheet before it is chilled.

Pressure Exerted by Squeeze Rolls

In the roll section, the bottom roll is of chilled iron, either solid, or bored with a hole for steam. During the operation, the iron roll, if solid, is heated by contact with the molten wax. The top roll is a composite, having an iron core with an inch of rubber for covering. The density of the rubber is determined by local conditions. In the newer machines the rolls are mechanically, rather than manually, adjusted and are held in position to compensate for the pressure of the wax at the nip. An air-operated loading device is used, with gages to measure and record the presure at both ends of the nip. The difference in pressure compensates for any variation across the dry sheet. With pressure indicators and recorders, the degree of pressure can be determined for each grade of waxing, kept substantially uniform throughout each run, and repeated on the next run of the same grade. In general, a tighter squeeze is necessary to keep the coating down to specifications at higher speeds. With machines that have no gages, there is a danger of using too much pressure and damaging the rubber roll.

In a three-roll squeeze section, the bottom roll is chilled iron and the two top rolls are hard-rubber covered.

Density of the Rubber Roll

In general, relatively soft rubber will apply more wax to the sheet than a harder covering.

Melting Point of the Wax

In dry-waxed sheets, a low melting point (under 130°F) grade is generally used as these grades are less expensive, and the melting point is not important where the wax is thoroughly impregnated. At the other end of the scale, glassine requires wax of the highest melting point to produce the hardest possible surface and best possible finish.

In the case of most wet-waxing processes, a wax of higher melting point may be used in summer than in winter to reduce danger of blocking in hot weather. Blocking tendency can be reduced by the addition of 15 to 20% microcrystalline wax to paraffin. (Above 20% the tackiness of the microcrystalline wax becomes effective.)

Temperature of the Wax Bath

The temperature of the wax bath should not be too high. For wet-waxing, 15 to 20°F above the melting point is recommended. Temperatures above 175 to 180°F for paraffin are not advised because of the danger of oxidation where the waxes may be continuously mixed with air through agitation.

Molten paraffin wax is a thin liquid, and the viscosity drop with temperature increase is only slight. Higher wax bath temperatures are therefore not so important in securing better penetration or decreasing the wax load. Microcrystalline waxes, which are more viscous than paraffin when molten, also show a greater rate of viscosity increase with decrease in temperature, so that for these waxes the temperature of the wax bath is a more important factor in determining degree of penetration and wax load in the sheet.

Accurate control of the temperature of the melted wax is important. For close control, one plant uses recording-controlling instruments, and two wax tanks for each machine. The first tank has the wax heated to the approximate temperature level, and the second tank, through which the paper passes, has the temperature maintained within one degree fluctuation. The temperature of the bath is regulated at the start of a run by setting the controls at the desired temperature. The wax baths are heated by steam coils or a steam jacket in the bottom of the tanks. Close control prevents overheating and oxidation of the wax.

Presence or Absence of Blankets

The wax load can be increased by the use of blankets. Newer blankets generally produce heavier wax loads. Blankets are not needed in the wet waxing of light weight sheets and cannot be used for dry waxing.

Chilling by Water Bath or Cold Rolls

For good gloss and opacity in the wax coating, the water bath is recommended. Water waxers chill the sheet quickly from both sides, and produce a coating consisting of small crystals and needles, which reflect light. The temperature of the cooling water should be between 32 and 40°F. Refrigeration systems are necessary to maintain this temperature range the year round.

The cold-roll machine is better suited for the manufacture of transparent types of paper. The lower cooling rate, first from one side of the sheet and then from the other, permits formation of large plate-type crystals (in the case of paraffin) which transmit light.

FURTHER FACTORS

Air Conditioning

One modern waxing plant has installed an air conditioning system to reduce the danger of having the wax block in the rolls in summer.

Heat Sealing

In some types of packaging, especially bread wraps, it is required that the waxed paper be self-sealing, by heat. The parts of the paper where the seal is to be made are pressed together and heated until the wax melts. The heating is discontinued, and the paper is held together until the wax solidifies, at which time the wax will have sealed the parts of the paper together. This seal is not strong with paraffin wax alone, but is usually strong enough for the purposes required. If a stronger bond is required, certain materials can be added to the wax bath, as previously described under "Wax Formulation," p. 221.

For good heat-sealing qualities, the surface wax film must be of sufficient thickness in relation to the finish of the paper. The paper must not absorb too much of the wax during the heating process or too little wax will remain to provide a seal. A high-finish sheet of low wax absorbency does not need as heavy a wax film as a lower-finish sheet of greater absorbency, but under the proper conditions, the latter produces a stronger seal.

Petrolatum and Oil Processes

Petrolatum is applied to paper in two ways. The first method is the same as that used for the dry waxing of paper, where the chill rolls are converted to hot rolls to produce thorough penetration of the impregnant. The second method is to apply the material at the calender stack of the paper machine either by using a special "water" box or by means of a bath below the stack. The roll in contact with the impregnant and the roll directly above act as squeeze rolls. Steam coils heat the bath to keep it fluid. One or more of the rolls of the calender stack can be heated to improve penetration.

Methods of applying oil are the same as for petrolatum. The oil bath can be unheated or warmed slightly to improve penetration. Reduction of the oil load in the paper can be brought about by the use of scraper bars arranged in the form of a comb. These bars remove part of the oil that is brought up from the bath by the bottom squeeze roll. The amount of oil can be controlled also by varying the width and number of the bars. The streaks in the sheet produced by this method can be eliminated by winding the finished roll tightly and allowing it to stand for 24 hours or more. The oil then becomes evenly and thoroughly distributed. The oil load in such cases can be as low as 5%. (Oil or petrolatum load in most cases is of the order of dry waxing, namely 15 to 20%.)

Laminated Papers

Laminated papers consist of two or more layers (plies) of paper, which have been combined into a single structure by means of an adhesive between their inner surfaces. Cost and other considerations, as a rule, limit the number of plies to four; two is the usual number. The plies can be of different types of paper, film, or foil so that the distinctive properties of each can be combined to give a product of wider usefulness. Where the adhesive composition or laminant used is wax, advantage is taken of its water-vapor-proofing qualities to produce a sheet of high water-vapor resistance, without the presence of wax on the surface.

The many combinations of paper and laminant fall into three main groups: (a) two hard-surfaced sheets, (b) a hard-surfaced and an open sheet, and (c) two open sheets. Group (a) consists usually of glassine, cellulose acetate film, or Cellophane, which are useful in food packaging because of their transparency or translucency. Group (b) includes glassine or parchment bonded to paperboard, kraft, or sulfite papers. The glassine or parchment is used for its greaseproofness. In this group are also combinations such as foils and sulfite or other papers. Group (c) generally includes light-weight sulfites or krafts laminated with wax to produce a water-vapor resistant sheet without wax on the surface.

The four main types of laminants are aqueous solutions, latices, solvent solutions, and hot melts. The last are waxy, asphaltic or resinous mixtures, of which the most important and most widely used is wax. The wax is generally of the microcrystalline type because of its tack, and its ductile and flexible nature. The wax bond is a continuous film, and so provides a combination with a high degree of water-vaporproofness. For some applications, the wax may be compounded with other materials. Many patents cover such mixtures, equipment for laminating, and methods of operation.

The wax bond must be strong enough to prevent easy separation of the plies or slipping of one ply on the other. In the case of a two-ply sheet, there are five possible separation points—namely, in either layer, at the bond between either of them and the wax, or in the wax layer itself. In the case of group (a), which requires a wax of high adhesiveness because the surfaces are hard and smooth, splitting would occur in the wax layer, if the sheet is properly bonded. With group (c), the opposite is true. Here high wax tensile strength rather than high adhesiveness is important, and rupture will normally occur in either paper layer. Group (b) is intermediate between the other two, and rupture will normally take place between the wax and the hard-surfaced paper.

Two basic rules have been given for laminating: (1) The wax should always be applied to the harder surface of the two sheets. If both are equal, the relative ease of handling would govern the choice. (2) Temperatures should be so adjusted that by the time the waxed sheet reaches the combining rolls, the wax is only slightly above its solidifying point. If the temperature is too low, the wax will not penetrate the unwaxed sheet sufficiently for a good bond. If the temperature is too high, too much penetration will result and the layers may spring apart after leaving the nip of the combining rolls. If the wax temperature is correctly controlled, contact of the waxed layer with the cold unwaxed layer serves to solidify the wax at the junction point of the plies. This means thermostatic control in the wax preheating tank and in the waxing pan. Heat is not applied at the combining rolls.

In the laminating operation, the underside of the harder-surfaced sheet is waxed by passing the sheet through the upper nip of a three-roll stack of waxing rolls. This method provides for more uniform application of the wax. The bottom roll revolves in the wax pan, transfers the wax to the middle roll, and this, in turn, transfers it in the form of a continuous film to the underside of the sheet. The waxed sheet then passes through the nip of one or more sets of combining rolls where it meets and is bonded to the other sheet, and finally passes to the winder. Good grades of paper should be used, especially in the case of the open-type sheet to provide greater coverage by the wax and more economical operation. Good grades of paper also result in the production of a sheet of higher general quality.

OTHER MEANS OF WAXING

The principal method of applying wax to paper is by means of waxing and laminating machines. There are other techniques in addition.

Beater Sizing

Wax emulsions compatible with pulp and rosin size are added to the pulp during the stock preparation. These emulsions are so formulated that they can be precipitated on the fibers upon the addition of alum to provide proper distribution and retention of the wax in the finished sheet.

Top Sizing

By this method the wax emulsion is applied to the surface of the sheet, usually in a size press or by means of water boxes on the calender stack during the manufacture of the paper.

Dacca Method

By the Dacca method empty cartons, glued at one end, are completely immersed in a molten bath compounded of pale crepe rubber or synthetic resins and a blend of waxes, the compound being tasteless, odorless, and nontoxic. The cartons are drained to proper weight under controlled temperature conditions and cooled.

Single Dip-Method

The cartons are first filled and sealed, then immersed once in the wax bath. The cartons must be completely sealed to prevent the flow of wax into the filled package.

Flushing or "Enrobing" Method

Filled and sealed packages are passed twice by sets of nozzles so arranged as to flush the packages of molten wax from all four sides. The packages are held so that the excess wax drains from one corner, rather than from one edge. The excess wax from the nozzles and packages drains to a sump tank where it is strained and then passed to the pump section and returned to the nozzles.

Double-Dip Method

The containers are first filled and sealed. On the first dip, the package is partially submerged to a controlled distance. The package is then inverted and given the second dip, which slightly overlaps the first.

APPLICATIONS

Waxed paper finds literally hundreds of applications. Practically every industry has some uses for this product. It can be in the form of transparent or opaque wraps either put directly around the article or around cartons holding the articles.

It can be employed as a liner for cartons, boxes, or bags; in laminated wrappers and bags; and as a separator, interleaving, slip sheets, or protective papers of various other types. It is sold either in rolls or in sheets, depending on the use for which it is intended.

The Waxed Paper Institute—the association of the waxed paper manufacturers—with headquarters in Chicago has compiled a list of the principal uses. The largest part of the waxed paper production is for the food industry. Most of this goes to bakeries, for bread, cakes, pies, cookies, crackers, and doughnuts. Butter and cheeses, oleomargarine, candies, meats and fish, quick-frozen foods of all kinds, vegetables, and yeast are other food products for which waxed paper wrapping of one form or another is used. Waterproofing paper for food packages is described by Heiss.[2]

The metal goods industries use waxed paper for wrapping automotive and machinery parts, razor blades, tin-plate products, and as separator sheets for batteries. The textile industry uses it for wrapping black friction tape, thread, rayon, and nylon. In hospitals, it is used for sanitary sheets, compresses, wraps for surgical specimens, and gauze bandages.

Waxed paper slip sheets find an important use in connection with printing, mimeographing, and decalcomanias. The United States Government Printing Office uses these sheets for stamps, both in book and sheet form.

Cake glue, soap, candles, chewing gum, plants and shrubs, rubber heels, tires, and tubes are other products that are wrapped in waxed paper. Trimmings and shreds are used for packing purposes. Water-vaporproof cartons, baking cups, household rolls, and soda straws are made from waxed paper or board.

Petrolatum- and oil-treated papers, or papers treated with a blend of these and wax, are used for meat wraps, dusting papers, metal-parts packaging papers, fruit wraps, "parchment" lamp shades, tympan papers, and paper specialities.

REFERENCES

1. Yozek, M., and Utschig, W. C., U.S. Patent 2,912,347, 1959.
2. Heiss, *Mod. Packaging*, **31**, 119-24 (1958).

chapter 10

Gummed and Specialty Tapes and Labels

F. W. FARRELL

The manufacture of gummed paper is one of the major converting processes. The base paper is bought in the form of rolls from the paper manufacturer and the adhesive is applied at the gumming plant. The basic function of the industry is to supply adhesives in a more useful and economical form than in the uncertain and wasteful manner of the olden days. The old glue pot and mucilage have given way to controlled adhesives applied to the proper paper, which is selected according to its specific use.

The gumming industry, as far as production is concerned, is relatively new, and its active expansion probably began around 1900.

TOBELMAN PROCESS

Two factors contributed to this expansion. First, the Tobelman process of flattening gummed papers; second, the introduction of the corrugated fiber shipping case. Prior to the Tobelman patent, it was almost impossible to print the curly stock produced in gumming. By the process disclosed in this patent, the paper can be flattened so that it is possible to feed it into the press. The flattening is accomplished by breaking or cracking the gum so that the tension between the gum coat and the paper is eliminated to a great extent. By proper humidification this flattened stock can be printed today on fast, automatically fed presses. The introduction of the corrugated box is responsible for the newest branch of the industry, because kraft-sealing tape is used in sealing these boxes or cases. In addition, the manufacturer's

237

joint for these cases consumes important tonnage of cloth-based gummed material.

The previously mentioned two events are also responsible for the growth of the two main branches of the gummed-paper industry. One branch is concerned with the manufacture of white stock for label use and, therefore, its product is mainly a sheeted one; the other branch is engaged in producing kraft base stock that is generally sold in the roll form for sealing purposes.

The preparation of the adhesive and the processes of gumming are much alike, but the two branches differ after the gumming operation. The label stock, if it is to be sheeted, is flattened, calendered, sheeted, sorted and packed, usually in a humidified atmosphere. The roll stock, for either label or sealing application, is flattened, calendered, slit into individual rolls, and then packed for shipment.

The machines used for applying the adhesives (Fig. 10-1) may be of the usual three-roll system, i.e. a lifting roll, a spreading roll, or an applicator roll; or they may be more complicated such as the reverse-roll coater.

The more elaborate and expensive coating methods and machines are not necessary for the gumming operation because of the natural

Fig. 10-1. Gumming machine (*Courtesy of John Waldron Corp.*).

mobility of the adhesive solutions that permits them to flow evenly after application.

As the adhesives are most often applied hot, the gumming machine has a jacketed glue pan heated either by hot water or by steam. For uniform and dependable results, these pans should be fitted with suitable thermoregulators.

The gummed sheet, as it leaves the machine is dried in a number of ways, but usually by passing through heated chambers. On account of the inherent curl produced when the gummed sheet dries, it has to be kept under tension. It may be supported on a carrier belt, passed over a revolving wheel, or it may be supported on rolls positioned at either end of a drying tunnel. A blast of hot air is applied to the gummed side and often means of humidity control are positioned near the wind-up unit.

The light weight of the white-label stock often requires a bare edge to offset the breakage produced by nicks in the edges of the sheet. Dry gummed paper is very brittle and does not withstand much tension.

Drying temperatures vary with the type of adhesive used and the speed of drying desired. The faster the machine operates, the higher the permissible heat, but the gummed stock should never be dried to less than 4.5% moisture. This is particularly true of glue-gummed sheets, which may otherwise lose the necessary rapidity of retack. When drying the glue-gummed stock, precautions should be taken that the outer layer of adhesive is not dried before the lower layer, or two detrimental effects may be produced: The steam or moisture escaping from below may cause bubbling or pitting; or when the stock is later held in the roll, the moisture will diffuse from the wet underlying layers and cause serious blocking. The equipment used should be supplied with the proper unwind and rewind assemblies, with tensioning controls.

ADHESIVES

The most common adhesives used in the gumming industry are glue and dextrin. Others used to a less degree are starch derivatives, water-soluble natural resins like gum arabic, and asphalt emulsions. The present trend is toward synthetic resins, modified starches, lignin and similar adhesives.

The glues used are made from either hide fleshings or animal bones. They are graded and sold on the basis of their viscosity or jell strength. In the gumming industry, hide glues of about 150-g jell strength and bone glues of about 70-g jell strength are used. The average viscosity, expressed as millipoises, is 70 for hide glue and 40 for bone glue. The process of manufacturing glue is a simple hydrolysis, the hide fleshings or bones being cooked with boiling water. After thoroughly cleaning the hides or bones they are cooked either in open or pressure tanks and the solutions are run off. After this first extraction, it is customary to make several further ones.

The first extraction yields the higher test glues. The products of additional extractions are of decreasingly lower quality. The extractions are mixed and then concentrated by evaporation under vacuum. The concentrated glue is run into pans and chilled. The resultant jell is sliced into slabs spread on wire trays. The resultant dried cakes are ground and packed in bags for shipment. In addition to the ground glue made as indicated, a high tonnage of the so-called "wheel" glue is produced. This is made by running the concentrated extract on the surface of a heated drum from which it is scraped off by a doctor blade after it is dried. Ground glue has a moisture content of 11 to 14% whereas "wheel" glue contains 8 to 12% moisture.

Dextrin is manufactured from starch by heating it in the dry form in steam-heated jacketed kettles. During the roasting process, the starch is constantly stirred. Usually the starch is sprayed or mixed with a small percentage of a volatile acid such as hydrochloric, nitric, or a mixture of both prior to the roasting operation. The grade of dextrin produced is governed by the time and temperature of roasting and the amount of acid used. By varying these factors, one can produce a very soluble product or one of low solubility. The lower-solubility products are usually of lighter color and, in fact, may be nearly pure white, whereas the more fully converted ones are tan to brown in color. Two other starch products are used to some extent, i.e. British gum and chlorinated or oxidized starches. When the conversion point is reached, the slurry is diluted, and the starch washed and then dried.

Of the natural water-soluble resins gum arabic is most commonly used. It occurs in the Mediterranean area as a resinous exudation of trees of the acacia family. The nodular masses are collected by the natives, sorted and graded as to color and freedom from dirt, sand, and bark, The most commonly used grade is known as grade No. 1, which is light tan in color and is sold in lump form.

Synthetic and natural rubbers and resins, lignin, asphalt emulsions, etc., are used in lesser amounts, and their preparation and compounding are mostly trade secrets or covered by patent applications at this time.

The resin- and rubber-base coatings are generally used on label papers and may be of the moisture-seal, heat-sealing, or pressure-sensitive types. Because these adhesives are more expensive than those used for gummed labels, they are often laid down on more expensive base sheets than the sulfite-soda base of the average gummed label. Aluminum foil, Cellophane, and even plastic films are used as well as white or colored label papers. Some of these specialized coatings are also used in the pressure-sensitive sealing-tape field to produce items such as Scotch tape.

PREPARATION OF THE GUMMING AND COATING MIX

The best method for dissolving animal glues is as follows:

A jacketed kettle is used. The jacket contains water heated either with live steam led into the water, or, preferably, the jacket is equipped with steam coils. Cold water is run into the kettle, the weight of cold water being the same as the weight of the glue to be used. The glue is slowly sifted into the water while the agitator is revolving so that no lumps of dry glue are formed. Agitation is continued for 10 to 15 minutes to wet all particles of glue; then the agitator is stopped and the glue permitted to soak for 45 to 60 minutes. Then steam is applied to the jacket and the agitator started again. The temperature of the glue solution is brought up to 140°F and kept there until the glue is entirely dissolved. Modifying ingredients are added at this point: fish glue, plasticizers, such as glycerin, glucose, invert sugar; preservatives; and scent. The mixture is agitated until all ingredients are incorporated; then the glue solution is ready to be drawn off through a strainer. It is important that the glue solution is not heated above 140°F as higher temperatures for prolonged periods of time will break down the protein molecule and lower the adhesive qualities of the glue.

To dissolve dextrins for flat gumming and as extenders for glue mixes, the same type of equipment is recommended. The procedure is almost the same, with the following modifications. Add only about 90% of the water to be used in the kettle, start the agitator, and sift the dextrin slowly into the water, breaking up the lumps as much as possible. After all the dextrin has been added to the water, part of the dextrin will usually float on the top of the water. Add the remaining water to it at this point so as to wet all the dextrin, continue agitating for about 15 minutes, and turn on the heat. To obtain best results, heat the solution to 185 to 190°F, turn off the heat, stop the agitator, and let the solution stand for 30 to 60 minutes with the lid tightly closed. This schedule will permit the foam to rise to the surface and break up almost completely. If modifiers are used, add them to the hot solution before stopping the agitator. If the dextrin is to be used as an extender, cool it to 140°F before adding it to the glue solution. If it is used for flat gumming, draw it off through a strainer into a storage tank and let it stand for at least 24, preferably 48, hours before using it on the machine.

To make sure that the gumming mixes have been prepared properly and that the gumming machines can be operated efficiently without the need of constant adjusting, viscosity and solid content standards should be established for each gumming mix. The viscosity can be checked by running the solutions through a Zahn or Ford cup or using a Brookfield rotational viscosimeter. Make these tests at the same temperature; a few degrees difference in temperature will make a considerable difference in the obtained results.

For the determination of the solid content, a hydrometer graduated in Baumé degrees can be used. Be sure that the cylinder in which the mix is placed is at least 6 in. in diameter so that the mix is not chilled too rapidly, and that the mix is drawn off the bottom of the kettle and not taken from the top, so that there will be practically no foam in the mix to be tested. The temperature again should be the same for every test. The hydrometer should be perfectly clean; it should be washed in lukewarm water after every test. The mix is placed in the test cylinder, which should be at least as tall as the hydrometer and the hydrometer is lowered into the mix gently until it comes to rest and the Baumé reading is recorded. If the hydrometer is dropped too rapidly into the mix so that it sinks and then rises again, the hydrometer should be cleaned and the test repeated.

A dry animal-glue film is quite brittle, especially if higher-jelly-strength glues are used in the preparation of the mix, and is apt to fly off the paper when it is broken or after the tape has been stored at low relative humidities for a longer time and the tape is used on high-speed machines. For this reason, a small percentage of plasticizer is used in the mix, about 2 to 4%, based on the dry weight of the animal glue. Glycerin or the glycols are best for this purpose, but for reasons of economy they can be partly replaced by glucose or invert sugar. When using invert sugar with animal glues, be sure that the grade is the one recommended by the manufacturer for this purpose; other grades contain aldehyde groups that tan the glue and make the tape unfit to use.

Fish glue imparts the quick tack needed for tapes used on high-speed, automatic machines. The quantity used depends on economies and the grades of glue used in the mix.

Glucose and invert sugar are used most generally for plasticizing flat-gummed dextrins. They reduce the brittleness of the glue film, which is important in the breaking operation, and render the finished sheet less apt to curl, i.e. they will extend the range of humidities at which the properly finished sheet lies flat. The percentages used are about 2 to 6, based on the dry weight of the dextrin.

Preservatives are used to keep the glue mixes from spoiling, especially if they are to be kept for a longer periods of time such as shutdowns over the weekend, or to prevent the tape from molding when it is used in the high humidities encountered in cold-storage plants. The most commonly used is zinc sulfate, which has been replaced largely by the chlorinated phenols such as Dowicide and Santobrite. The percentages used are small, less than 1%, as some individuals are very sensitive to these chemicals, and break out in rashes on handling them.

Dextrins and extenders are employed for two reasons: About 10% mixed with the animal glue produces a tape that is wetted and becomes highly adhesive in a shorter time. The other reason is economy when dextrins are cheaper than animal glues, or when glues are scarce on the market. A British gum-type extender is the best kind of dextrin to be used with animal glues, and the percentages depend on the kind of British gum obtainable. Tapioca British gums are better adhesives than the ones made from corn; the same applies to flat-gumming dextrins.

Scents are used to mask the odor of animal glue and impart a better odor to labels with dextrin gumming. The most commonly used is methyl salicylate or oil of wintergreen; other scents are essential oils, like peppermint, sassafras, and cinnamon. Their use depends largely on the manufacturer's taste and preference. The quantities are very small, usually fractions of 1%.

A number of other chemicals are employed to obtain papers for particular applications, but their use is usually the manufacturer's secret, with the exception of the patented ones.

MANUFACTURE OF GUMMED TAPES AND LABELS

Gummed Flat-Label Papers

Gummed flat-label paper is used for all types of stamps, labels of innumerable types, pennants, posters, seals, window stickers, and many other such common items. Certain types of resinous materials are used for labels, and their development has been rapid. Their use at present is generally limited to special cases because of cost, but the water-insoluble adhesives will probably become increasingly popular. Many of the details of the water-soluble gummed paper to be discussed in this section also pertain to the water-insoluble type adhesives.

Raw Stock

Many types of paper are gummed for use in the label trade, including coated paper of various colors and grades, such as metallic coatings of gold, silver, and bronze, as well as the plated and glazed sheets. Also used are the uncoated sheets, which include the various natural shade and colored krafts, mediums, and many papers of machine finish, English finish, and high-finish whites. In addition, water-soluble adhesives may be applied to metal-foil papers, glassine, Cellophane and the plastic films.

Weights

Even though sealing tapes are not covered in this section, various weights of brown, white, and colored krafts, gummed and processed successfully for the label trade are worth mention. Weights employed for labels in the kraft line range from 35 to 60 lb (24×36—500). Weights usually vary, for coated or uncoated paper, from 16 to 30 lb (17×22—500). Both lighter and heavier weight sheets are sometimes gummed for special purposes in both of the classes mentioned.

Specifications

The following specifications should be met to insure proper raw stock for gumming and processing. The paper should be properly sized to the extent that the adhesive used for gumming purposes does not penetrate unduly into the sheet. A poorly sized sheet will obviously require more adhesive which will of course, increase manufacturing costs. A surface similar to that of a good printing sheet is usually acceptable as a good gumming surface. The sheet should also have a satisfactory printing and writing surface, e.g. a coated surface, for most of the gummed paper is used for printing purposes on the plain side. The paper should be smooth and free from dirt. Creases cause considerable waste in the manufacture of gummed papers, as they commonly do in other uses of paper. Uniformity of thickness should be rigidly met, certainly in many specialty items. This precaution is particularly necessary because in the flattening process, uneven gages cause slack edges that contribute to misgumming as well as irregularity of tension in the drying operation, which possibly could lead to a curled sheet. These slack edges would also prevent the manufacturer from obtaining the same degree of flatness across the sheet. Paper structure or formation is extremely important, as a sheet that shows a wild formation (i.e. irregular structure) does not absorb the adhesives properly or evenly.

The tests of raw stock for a gummed paper are the same as the regular methods used for evaluating a sheet for ordinary coating work. The most common tests determine thickness, weight, tensile, and tearing strength. Dennison wax tests are made along with other sizing tests to be certain of proper printing and writing surface.

Gumming

Common adhesives are animal and fish glues, dextrins, and water-soluble natural gums or combinations of these. For ordinary flat-gummed papers, the amount of adhesive usually applied is 6 lb (17×22—500) on the dry basis. For specialties, when the paper is heavy, the amount of adhesive used will reach as high as 10 lb (17×22—500).

After the adhesive has been prepared in the mixing room, the proper temperatures must be maintained for correct application by using a glue pan that is steam- or hot-water jacketed. The correct working temperature is usually about 125°F. The adhesive is applied to the paper by the kiss- or squeeze-roll method in most mills. For most types of work a 40%-solids adhesive is satisfactory, but the solids content varies with the type of raw stock or the type of adhesive. For example, a soft-sized paper should be gummed with an adhesive of a higher solids content than that for a hard-sized raw stock.

After the gumming operation has been completed, the next step is the flattening. Most of this type of paper is sheeted and later printed. For obvious reasons, these sheets cannot have the curl that results from the application of the adhesive. Gummed papers curl to the gum when the moisture content is low, and toward the plain side when the moisture content is increased because the gummed side is more hygroscopic and has a greater coefficient of expansion and contraction than has the plain side of the paper. The film of gum must be properly broken to make the sheet flat enough to be fed into the printing presses

Fig. 10-2. Gummed-paper flattening machine (*Courtesy of John Waldron Corp.*).

under variable conditions of humidity at the high rate of speed at which these machines operate.

A roll of gummed paper, with the gummed side out, is passed over a bar under tension at a 45° angle. It is then necessary to pass the sheet over another bar, making a 90° angle with the first bar. Without the second bar, the sheet would curl from one end to the other. The second bar obviously removes this curl.

In the flattening operation care must be taken (Fig. 10-2) that the ungummed side of the sheet is not roughed up and thus mars the printing surface. Caution should be taken to prevent the flaking of the glue film.

The typical finishing operations are similar to those used in finishing ordinary coated paper. Machine-finished papers are usually calendered, depending upon the customer's requirements, and upon the finish desired. The regular multiple cutters and trimmers are used for the remainder of the finishing.

Gummed sheets are usually tested for sticking qualities, hygroscopicity, flatness and writing quality.

Sticking quality of the heavier grades of paper is measured by the

McLaurin tester. This measures the tack of the gummed paper after it has been moistened, and before it is dried. Other gummed papers are tested by moistening one-inch strips and applying them to the type of surface for which the adhesive is intended.

Hygroscopicity is measured by stacking 2-in. squares of gummed paper and observing their action under the same pressure in various ranges of controlled relative humidities. Standard tests for blocking are given in Federal Specifications for Blocking UU-T101 A, March 3, 1939, and subsequent amendments.

Flatness is tested by placing the sheet in a room under controlled conditions and observing its behavior. Writing quality is determined by a pen-and-ink test.

Odor, grease, and color of the adhesive are important, but are controlled in the adhesive preparation as well as in the gumming operation. Inspection should be made of the finished paper to insure that it is free from grease and odor and that the color of the adhesive is correct.

The method of sorting and inspecting gummed paper is akin to that used in ordinary coating and paper mills. Sheets are inspected, and all that are dirty or below standard quality for the usual reasons are separated and placed with seconds. Both the plain and the gummed side are inspected because, as stated previously, the plain side must have a satisfactory printing surface.

After the paper has been inspected and all imperfects are removed, it is trimmed to size (usually 17×22—500 or 20×25—500). It is wrapped in waterproofed paper to protect the adhesive during shipment and storage. The utmost care should be taken in wrapping this type of paper, as the adhesive would cause blocking when stored in highly humid atmosphere. Obviously this is especially true during the summer months.

Conditions of storage rooms are extremely important for this kind of paper. Every year considerable losses occur because of the failure of paper and printing houses to store gummed paper properly. Paper of this nature should never be placed near radiators, or any type of heating unit, nor near windows or openings that may allow rain or dampness to penetrate into the room. Once a package of gummed paper is opened to remove only part of the contents, the package should be securely rewrapped to prevent the paper from taking up moisture.

Sealing Tapes

Sealing tape is made in three categories: light, medium, and heavy weight. Light-weight sealing tape is designed generally for light-weight bundles and bags, such as those put out by retail stores. Medium-weight tapes are employed most extensively for sealing the flaps of corrugated or fiber cartons. Heavy-weight tapes are designed for heavier packages and cartons.

Because strength is a prime requisite in such a tape, the backing material is almost wholly of kraft, which may be brown (or natural shade), or white (bleached) or any of a wide variety of colored kraft papers. It may be printed or plain.

For light-weight sealing tape, a basis weight of 35 to 45 lb (24 × 36 —500) is employed. For medium weight, a 60-lb paper is standard. For heavy weight, a paper of 90 to 120 lb is common.

Because sealing tapes are usually cut to narrow widths, the width of raw stock roll is limited only by the capacity of the gumming machine and the economical utilization of the paper machine capacity in width. The diameter of the roll is dependent on space for handling and ease of handling at the gumming machine. The diameter may run from 24 to 36 in.

The paper should have the strength of a No. 1 grade kraft in bursting, tensile and tear resistance. In uniformity it should conform to recognized trade customs. The paper should be of good density, uniform in thickness. It should be adequately sized, and at least one face should be smooth, suitable to maintain an effective and economical film of glue.

Adhesives are largely of the animal glue type, applied hot, and as concentrated as possible. The amount of adhesives (dry basis) runs from 12 to 15 lb/ream (24 × 36—500) for light-weight tapes, from 15 to 18 lb/ream for medium tapes, and from 17 to 20 lb for heavy-weight tapes.

Gumming speed is limited principally by the rate of drying, which, in turn, depends on dryer capacity, amount of glue, proportion of water and the nature of the adhesive. Speeds of 200 to 600 ft/min are employed.

The dried gummed paper passes to a rewinding mechanism where it is wound usually in a large roll corresponding to the raw-stock roll used for the gumming operation. The rewinder here is usually of the center-wind type and friction driven.

Slitting may be done on two general types of machine—shear-cut and score-cut. The first is equipped with a pair of circular knives for each cut; these overlap and cut much as do a pair of shears. The second machine has a single circular knife that cuts by pressure, the knife bearing on the gummed paper as the paper is carried partly around a cylinder.

The large roll of gummed tape, usually in the width in which it has been gummed, is slit, commonly in one operation, to a series of narrow coils of the specified width.

Directly from the slitting knives, the gummed paper (now in narrow tape form) passes to the rewinding mechanism. This again is of two general types. One is the center-winding type where a shaft-driven core winds up the tape. The second is the surface winder, in which the winding tension is frictionally imparted from a driven roll that rests directly on the winding coils of tape. Winding improvement is sometimes effected by a combination of center and surface winding.

The width of light-weight tape runs from ¾ in. upward, that of medium-weight tape from 1 in. upward, and that of heavy-weight tape from 2 in. upward, 1½ in. being perhaps the most common width for light weight, and 3 in. the most common width for medium- and heavy-weight tapes. Coils are usually packaged in bundles approximately 30 in. in diameter. Light-weight tape is wound in 500-and 800-ft. coils, medium weight usually in 600-ft coils, and heavy weight in 375-ft coils.

Since remoistenable gummings are to a certain degree hygroscopic, they are affected by change in humidity. They are more difficult to moisten in excessively dry weather and tend to block in excessively moist weather. For this reason, the coils are customarily double-wrapped, with a moisture-resistant waxed kraft inside; over this is the usual strong kraft wrapper. This arrangement assures delivery of a tape that is uniform and of unimpaired quality. Tape removed from the original bundle should be stored away from excessive heat or moisture, and preferably wrapped again in protective paper.

The finished product must be suitably strong, commensurate with the raw paper employed. Because it may be subjected to splitting or tearing strain in the length as well as in the cross direction, it should be made from paper of adequate strength in both dimensions.

The adherence of the tape, after suitably moistening and applying, should be such that upon drying and attempting removal, any rupture will be in the paper, either of the tape backing or of the surface to which the tape was applied.

Stay Tape

Stay tape is employed for the corner sealing of "set-up" boxes. In set-up box manufacture, the box blank, usually of chip board, is die-cut to size with the corners cut away so that after scoring and folding the sides meet with no corner overlap. Stay tape is used to join together at each corner the four sides of the shaped box.

A section of tape in length equal to the height of the box (or cover) is automatically moistened, cut and applied to the box corner. Single stay machines apply the four corner pieces successively; quad stay machines apply the four corner pieces simultaneously.

A No. 1 grade kraft is required. Since splitting or tearing strain, after application to the box, occurs parallel to the length of the tape, the paper should be sufficiently strong in this dimension. A paper in which the fibers lie predominantly lengthwise is inferior for this application.

The common brown kraft shade and gray (to roughly match the chip board) constitute the colors generally supplied for stay usage.

The raw paper is relatively heavy. In most cases, weights of 90 to 105 lb (24×36—500) are employed.

Like sealing tape, this stock is eventually slit to narrow widths, so that the raw stock width is limited only by the gummer width and the practical utilization of the paper machine capacity in width. The usual range of raw stock widths is 36 to 48 in.

The roll diameter, as for sealing tape manufacture, may run from 24 to 36 in.

The strength of a No. 1 grade kraft is required in bursting, tension and tear resistance. In uniformity, it should conform to recognized trade customs.

Minimum thickness for a given weight is important; the box manufacturer does not want a thick patch over his box corners. It should be well-sized and smooth, as for tape gumming, to maintain an effective and economical film of glue.

The adhesive for stay tape is similar to that employed for gummed sealing tape. As short sections of tape are sometimes employed (for instance, on a shallow box or box cover) the feed of moistened tape may be very slow and the glue formulation is generally adjusted to yield a relatively long period of tackiness after moistening.

For stay use, a glue coat of 17 to 20 lbs/ream (24 × 36—500) is desirable.

As the adhesive is similar to that employed for sealing tape, the gumming of stay tape should not differ greatly from that of gumming heavy-weight sealing tape.

For stay use, a clean, nonfuzzy edge is required and shearcut slitting is favored. Well-wound coils are also required, and a combination center and surface winder is preferred.

Stay tape runs usually from ¾ in. to 1 in. in width and is wound in coils of 10-in. diameter with paper cores having a 1¼-in. inside diameter.

This tape, like sealing tape, is usually double-wrapped with moisture-protective paper inside and strong kraft for an outer wrap. It should likewise be protected from excessive heat or moisture. It should be strong, commensurate with the raw paper employed, with special regard to resistance to lengthwise splitting or tearing. The adherence of the tape after drying should be such that any rupture, on attempted removal, will lie in the box board.

The adhesive should be of strong tack to prevent slippage from the box corner while the tape is still moist. It should be of sufficiently long tacky range so that on shallow boxes or covers, where tape feed is slow, the adhesive will not become too dry for good application.

Filled Cloth Tape

Filled cloth tape, generally in a kraft color, is employed both for the corner sealing of cartons and for diverse uses where a cloth instead of a paper gummed sealing tape is advantageous. In the field of carton corner sealing, filled cloth tape is largely superseded by tapes with other types of backing material.

The backing material is supplied by cloth processors as a clayfilled, starch-sized cloth. The type of cloth and amount of filling may vary rather widely, but in general it should be of suitable strength in each dimension and of fair flexibility, but should not be too limp. In weight, the filled cloth may vary for different requirements from 75 to 150 lb/ream (24 × 35—500).

The raw cloth generally employed by the cloth processors is 40 in. wide. Shrinkage involved in filling brings the width to approximately 38 in. for the gummer's use. The roll diameter, as for paper tape manufacture, runs from 24 to 36 in.

Specifications vary widely. In general, the user should see that the advantages of cloth over paper are suitably maintained in this filled-cloth product. Of these, tearing resistance in each dimension and flexibility are important. The filling should be adherent and should provide a dense foundation for the adhesive that is to be applied without open areas. It should not be unduly absorbent or softened upon the application of the adhesive.

Because this material still retains much of the roughness and unevenness of a cloth product, much greater amounts of adhesive must be employed than with paper; 35 to 40 lb/ream are customary, usually in two successive applications. The adhesive is usually an animal glue of greater strength than is required for paper tapes. The gumming operation is in general similar to that for gummed paper, but usually it is carried out at lower speeds as the cloth-backed tape is more difficult to dry. In general, slitting may be satisfactorily done on the equipment used for gummed-paper sealing tape. Coils are usually 1000 ft in length, the lighter weights in a 2-in. width and the heavier weights from 2 to 3 in. in width. Cores are of paper with 1⅝ in, inside diameter. The coils are packaged in cartons, usually of 6 to 9 coils, depending on the width of coil.

The finished product should be strong, commensurate with the filled cloth employed, with special regard to tearing resistance. It should be flexible, but at the same time stiff enough for feeding on automatic equipment.

The adhesive should be sufficient to compensate for the cloth roughness and to provide a smooth layer. It should be of strong tack and of fairly long tacky range to insure "grab" consistent with the more severe requirements for cloth tape. The adherence after application and drying should be such that failure, if it occurs, will not lie in the glue layer or in too ready separation of the filling from the cloth.

Duplex backed tape, a lamination of cloth and paper, is used extensively by carton manufacturers for sealing the open corner left after a corrugated or solid fiber carton is die-cut, scored and folded flat for shipment. Laminating is usually performed by the gummed tape manufacturer, the most common laminant being a modified-starch adhesive. The paper is a relatively light weight kraft 20 to 30 lb/ream $(24 \times 36$—500). It should be strong in both dimensions and resistant to splitting apart from the paper layer.

The cloth is usually of two varieties: (1) sheeting of evenly balanced strength in the long and cross dimensions is used for the lighter-weight product, and (2) a cloth called Osnaburg of heavier construction, with the fill (or cross) threads stronger than the warp (or long) threads is especially suited for the manufacturer's sealing of the carton corner.

A thread count of the order of 36 warp threads and 40 fill threads per inch is typical of the lighter weight sheeting. The weight of this material corresponds to about 5.55 yd/lb in a 40-in. width. This is roughly equivalent to 54 lb/ream (24×36—500). This cloth is usually marketed in 40-and 46-in. widths.

For the heavier Osnaburg goods, a count of the order of 32 warp threads and 26 fill threads per inch is typical. These fill threads, while fewer, are considerably heavier than the warp threads. The weight of this material corresponds to about 3.65 yd/lb in a 40-in. width. This is roughly equivalent to 82 lb/ream (24×36—500). This grade is usually 40 in. wide.

The laminant may be of animal glue or modified starch or any other suitable adhesive. It is essential in laminating that a balance in flexibility be maintained. The flexibility of the cloth must not be lost, but a sufficient degree of stiffness must be imparted so that the finished tape may be fed readily on automatic moistening equipment. From 15 to 25 lb of laminant per ream (24×36—500) may be employed, depending on the nature of the laminant.

The laminated product varies (on a ream basis) from 100 lb for the lighter grade to 140 lbs for the heavier grade. The 40-in. cloth shrinks to about 38 in. width in processing. A 46-in. cloth shrinks to about 44 in.

Roll diameter, as for other forms of tape gumming, may run from 24 to 36 in. depending on space and facilities for handling.

A good laminated product should be of suitable strength, particularly with respect to tearing in the longitudinal direction. For uniformity of adherence and economy of gumming, the cloth should be even, free from bunches and excessive variation in thickness. The product should be resistant to delamination or weakening of the laminating bond under the tension and wetting involved in gumming. The paper surface (to which the adhesive is applied) should be suitably smooth. Thin material is more economical of adhesive than is filled cloth. A coating of 20 to 25 lb/ream is customary and is usually applied in one gumming operation. The commonly used adhesive is an animal glue and of greater strength than that required for paper tape. The gumming operation is similar to that for gummed paper, but, owing to the thicker backing material and the generally greater amounts of adhesive, lower speeds are customary. In general, slitting may be satisfactorily done on the equipment used for gummed-paper sealing tape. Coils are usually 1000 ft in length, the lighter weights in 2-in. width and the heavier weights in 2-to 3-in. widths. Cores are of paper with an inside diameter of $1\frac{15}{16}$ in. The coils are packaged in cartons, usually of 6 to 9 coils, depending on the width of tape.

The finished product should be strong, commensurate with the cloth employed, with special regard to tearing resistance in the longitudinal direction. It should be flexible, but at the same time stiff enough for feeding on automatic equipment.

The adhesive should be sufficient to compensate for the roughness of the laminated product and to provide a smooth layer. It should be of strong tack. The adherence after application and drying should be such that failure would lie in the cloth or the carton material and not in the glue or in the cloth-paper lamination.

Sisalkraft, a patented product, comprises a 50- or 60-lb kraft backing $(24 \times 36—500)$ and a 30-lb kraft face, laminated with asphalt in which sisal fibers are distributed. For carton manufacturers' use as a corner tape these sisal fibers lie crosswise; for other uses, *Sisalkraft* can be obtained with the sisal fibers lengthwise.

This material, as sold to gummed tape manufacturers, has a total ream weight of about 200 lb. It is supplied in widths of $40\frac{5}{8}$ and $42\frac{5}{8}$ in. and is generally sold in rolls of a diameter of 26 or 28 in.

For carton corner sealing, as noted, *Sisalkraft* is supplied with the sisal fibers lying crosswise. This material is relatively weak to lengthwise strain, but such strain is not a factor in this specific application. Against crosswise strains, the product is so exceptionally strong that the only concern of the tape manufacturer is that the sisal fibers are well bonded by the asphalt so that failure will not be due to fiber slippage.

The asphalt employed should be tough and so compounded as to avoid excessive brittleness, especially in cold weather. There should be no seepage of the asphalt at temperatures to be expected during the gumming operation or on long standing under summer conditions.

Since the raw stock, with its sisal fibers, presents a rather uneven surface, it is necessary to employ 25 to 30 lb of glue per ream to assure an adequate adhesive surface. An animal glue of high strength is suitable. The gumming operation in general is similar to that for gumming paper. In drying, the material should be kept below the critical temperature of the asphalt (about 200°F) and for this reason, the rate of drying is lower than is common with gummed-paper' sealing tape.

After gumming, the sheet is usually embossed in a fine, all-over pattern.

All-over patterns are advantageous for several reasons. Under the embossing pressure any local areas that have separated during gumming can be relaminated. Surface distortion from gumming and from the underlying sisal fibers is minimized and a useful flexibility is imparted to the product. With this backing material, slitting is more difficult than with the ordinary paper tapes. A special rewinding mechanism, with a rubber-surfaced drum rewinder, is recommended. The tape width is 2 to 3 in. and the product is usually marketed in coils of 1000 to 2000 ft. The cores are of paper with an inside diameter of $1\frac{15}{16}$ in. The coils are packaged in cartons containing 2 to 6 coils. To prevent sticking from possible exudation of asphalt, the coils are separated by waxed paper.

No delaminated areas should appear in the finished product. The sisal fibers should be suitably spaced to provide a nonbunchy product of even resistance to tearing in the length. It should be flexible, but at the same time stiff enough for feeding on automatic equipment.

The adhesive should be sufficient to compensate for the uneven surface of the backing and to provide a smooth adhesive surface. It should be of strong tack. The adherence, after application and drying, should be such that failure would lie in the carton material and not in the glue or in slippage of the sisal fibers.

GUMMED HOLLAND TAPES

Holland cloth is a starch-sized, pigment-filled product, mechanically glazed to a high finish on one side. It is produced in a variety of colors and is characterized by its finish and density and by absence of the openness and roughness normally inherent in a cloth. Its principal use is in binding, such as for paper tablets, note books and passepartout.

Two general grades are prevalent, generally known as No. 1 and No. 2. No. 1 is the better grade and is characterized by a higher thread count, a closer weave and a lesser amount of filling. The finished weight of Holland cloths varies roughly, from 60 to 90 lb/ream (24 ×36—500). Holland cloths are marketed in 36- and 42-in. widths, and are usually put up in 2500-yd rolls.

No precise specifications are available. The thread count and weight, amount of sizing and pigment are largely in the hands of the cloth finisher. The most important qualities are appearance, smoothness, and the adherence of the filling. In this latter respect No. 1 grade is pronouncedly superior to No. 2.

The gumming of this material differs little from that of gummed paper tape. About 20 lb of adhesive per ream are applied, the adhesive being usually an animal glue of a grade similar to that employed for gummed-paper sealing tape. Gumming may be applied either to the glossy face or to the dullfinished back of the Holland cloth, depending on the finish the user desires on the ungummed side.

Slitting is not significantly different from the method employed for gummed-paper sealing tapes, a shear-cut slitter being preferred. Coils are 150 to 300 yd in length and $\frac{3}{4}$ in. or greater in width. Wooden cores of $\frac{9}{16}$-in. inside diameter are employed. The coil diameter may run from 7 to 10 in. The coils are packaged in bundles of approximately 28 in. in diameter with a waxed paper inner wrap and a kraft outer wrap. This bundle is usually shipped in a carton.

Appearance of the finished product is important and the filling should not loosen in handling or in applying the moistened tape. The adhesive should be of good tack, and sufficient to provide good adherence.

USES OF GUMMED PAPERS

Gummed papers are used for labels, stickers, sealing tape for corrugated shipping cases and packages, stay for the corners of set-up boxes, veneer, and tape tablet binding.

The largest tonnage is used in the label stock and a sealing tape. The label stocks are sold to the printers. The standard sizes are 17×22 in. and 20×25 in. and the quotations are based on these sizes. Most of this stock is white, although there is of course considerable demand for colored stock either as coated or as plated or medium papers. As has been pointed out, gummed papers can be printed on fast, automatically-fed presses so that the requirements as for the put-up and flatness are the same as with any stock that is to be printed.

Although the label stock is processed to be substantially flat, it is nevertheless more susceptible to humidity changes than are plain papers and a wide spread of relative humidity produces curling and misregister.

In a dry atmosphere the curl will be to the gummed side and in a high humidity it will be to the plain or "back" side. If several impressions are made, as in multicolor printing, care must be taken to have the atmospheric conditions the same each time the sheet is printed.

Sealing tape is used in packaging for the closure of corrugated boxes. The moistened tape is applied to the closed flaps and along the edges of the flaps. As corrugated cases are subject to the rules of the Interstate Commerce Commission, the proper procedure for applying the tape should be strictly observed. Band sealers (Fig. 10-3) are in common use.

Fig. 10-3. Container band sealer (*Courtesy of Vertex Company*).

METHODS OF APPLYING TAPES AND LABELS

From the time sealing tape was first produced, moistening machines have been provided for its use. The original devices were not much more than moistened felts, sponges, or rolls, with a saw-toothed edge provided to sever the tape. From these primitive methods, we have now advanced to the automatic machines, which deliver predetermined lengths uniformly moistened and clearly cut off. These machines are provided with means for warming the water supply. Although the machines are relatively costly, they are vital to economical results, both as to speed of operation and satisfactory bondage of the tape.

The stay tape for the corners of set-up boxes is applied by machines made for this purpose. This tape is supplied in rolls of ¾- or ⅞-in. widths and 9-in. diameter. Two types of machines, the single stayer and the quad, are common; as the names indicate, the first applies the stay on one corner at a time and the second serves all four corners simultaneously.

The choice of the type of machine used is usually governed by the depth of the box. The deeper boxes, which require a longer stay, are usually run on the single stayer, and the shallower ones on the quads. The "throw" on the single stayers may go up to 5 or 6 in. whereas that on the quads may be as low as ¼ in. A shoe box, for instance, would be run on a single stayer, whereas a handkerchief box would be run on a quad. Often the covers of shoe boxes requiring a shorter "throw," or shorter length of tape, are run on quads.

Veneer tape is applied by taping machines that make possible the assemblage of many strips of veneer into large sheets such as would be needed for desk tops and furniture panels. The veneer strips are accurately cut so that adjacent edges are butted true and even and run under two toed-in knurled rolls, one on each side of the joints. The moistened tape is rolled on as the veneer is held between the pressure rolls. The several sheets thus made are applied to the core stick upon which the proper adhesive has been spread; this assembly, together with others, is then placed on the platens of a veneer press. Pressure, and often heat, is applied to get the necessary bond.

The gummed stock used for the manufacturer's joint of a corrugated box may be paper on cloth. The rolls are 2 in. in width and up to 24 in. in diameter. The cloth tapes are usually made from various gray goods and are obtained by the mill suitably filled and finished to give a good smooth gummed surface and in rolls of over 2000 yd up to 40 in. wide. As the strain on the box corner is at right angles to the edge, the cross or filler threads should be the stronger. Therefore, the tear strength of the tape in the running direction, or lengthwise of the tape, is the most important. Depending on the requirements of the box to be taped, a light or a heavy construction is used, and in the trade two grades are commonly available. For exceptionally heavy duty, extra heavy or strong backing materials are usual.

As to the paper tapes used on the manufacturer's joint, some heavier-weight kraft and case-liner papers are employed in quantity; for shipping containers to be used in interstate commerce, these grades are not acceptable. For this purpose, it is customary to use sisal tape. This product is made by combining a 30-lb and a 40-lb kraft sheet with asphalt in which raw sisal fibers are embedded across the tape. This arrangement results in a less expensive tape than cloth, and one

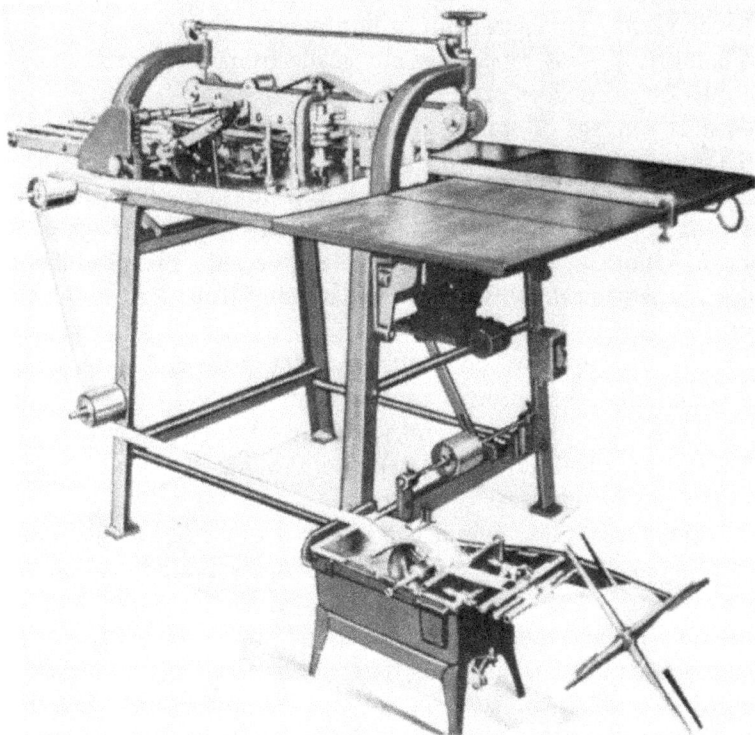

Fig. 10-4. Tablet stripping machine (*Courtesy of John J. Pleger Company*).

that is strong. It is not as pliable as cloth and therefore requires more care in application.

This tape is most commonly applied to the boxes on automatic taping machines. The corrugated boards cut to box size, creased, slotted, and folded flat so that the edges of the joint to be taped are adjoining and on the upper side are stacked in the magazine of the machine between two belts that pull the folded box forward. At the proper moment the tape is automatically reversed. The belts are about 12 to 18 ft long and 5 in. wide. They are maintained under substantial pressure.

Tablet stripping or binding uses coils of gummed material in a width of 1 in. and diameters of 7 to 10 in. There are various machines (Fig. 10-4) available for this purpose, and they all use an unusually large throw or length of moistened tape before adhering the stock to which they are applied. Illustration of the type of product includes school and stenographers' notebooks and pamphlets.

Much sealing tape is used in sealing parcels and packages in stores and laundries. This tape is usually made from 35-lb kraft in colors and may be printed or plain.

A large tonnage of sealing tape is printed. Most of this is printed at the gumming plant. However, some is printed by sealing tape jobbers both in machine rolls or trimmed rolls and in coil form. At the gumming mill, both methods are followed, but the machine roll method is both economical and efficient, because the production is higher and also because there is less danger of trouble from ink being offset on the back.

It is, of course, more economical to print in the coil form on small orders and these are printed on the gummed stock. Care must be exercised in using the proper inks so that offset is minimized, otherwise the oily ink will seriously waterproof the adhesive, a prolific source of complaint in this class of goods.

In most cases, these gummed products go to other manufacturers. Very few (e.g. some sealing tapes) are used by the general public in office or home. They are handled usually, if not always, by automatic machines. In fact, much sealing tape, and more each year, is applied by automatic moisteners such as the "tape shooter," which pushes out or throws a predetermined length of tape and at the same time moistens and cuts it.

chapter 11

Pressure-Sensitive and Release Papers

Pressure-Sensitive Papers

GERALD COLE

Since 1933, when pressure-sensitive paper, a new and revolutionary kind of adhesive coated paper, first came into being, a vital industry has developed until today world-wide volume has reached a point well above the $200 million mark, with plants and facilities in practically every country in the civilized world.

WHAT IS A PRESSURE-SENSITIVE PAPER?

Basically it is a laminate of three layers:

Layer No. 1—a substrate suitable for printing, i.e. paper, film, foil or textile.

Layer No. 2—an ever-tacky layer of either a natural or synthetic rubber material usually about 0.0007 in. thick, protected by

Layer No. 3—a release paper.

In general there are three types of adhesive coated papers: (1) Water- or solvent-activated (the gummed papers); (2) heat-activated; and (3) pressure-activated (or pressure-sensitive). Let us furthermore not confuse pressure-sensitive papers with pressure sensitive tapes, such as cellophane tape, surgical tape, and masking tape, Although the latter require most of the same types of adhesives, they do not use a release paper but are self-wound and can only be delivered in roll form.

Pressure sensitive paper, being protected by an adhesive coated release liner, may be delivered in any form: rolls, sheets or even fan-folded.

Because some of the terminology may be unfamiliar, the following glossary will be of help in further understanding this section.

Adhesion—Adherence. A bond established upon contact between two surfaces.

Adhesion, peel—Adhesion. Adhesion Strength. Peel adhesion is the force required to remove a pressure sensitive label from a standard test panel at a specified angle and speed after the label has been applied to the test panel under definite conditions. (Pressure Sensitive Tape Council.)

Adhesion, shear—Holding Power. The time required under specified test conditions to slide a standard area of pressure-sensitive label from a standard flat surface in a direction parallel to the surface.

Adhesive, cold temperature—An adhesive that enables a pressure-sensitive label to adhere or stick well when applied to a cold substrate.

Adhesive, permanent—An adhesive characterized by having relatively high ultimate adhesion.

Adhesive, pressure-sensitive—A type of adhesive which in dry (solvent free) form is aggressively and permanently tacky at room temperature and firmly adheres to a variety of dissimilar surfaces upon mere contact, without the need of more than finger or hand pressure (Pressure Sensitive Tape Council).

Adhesive, removable—A pressure-sensitive adhesive characterized by relatively high cohesive strength and low ultimate adhesion.

Barrier coat—Sealer Coat. A coating applied to the face material on the side opposite to the printing surface to provide increased opacity to the face material to prevent migration between adhesive and the face material.

Anchor coat—Primer. Tie Coat. A coating applied to the face material, prior to the application of a pressure-sensitive adhesive, to improve anchorage.

Caliper—Thickness. The thickness (as of a sheet of paper) measured under specified conditions. It is usually expressed in thousandths of an inch (mils or points). (*Dictionary of Paper*).

Cast-coated paper—Gloss Paper. A paper, the coating of which is allowed to harden or set while in contact with a finished casting surface. Cast-coated papers have, in general, a high gloss. (*Dictionary of Paper*.)

Cold flow—Ooze. The viscous flow of a pressure-sensitive adhesive under stress.

Die—Any of various tools or devices used for imparting or cutting a desired shape, form or finish to or from a material.

Die-cut—(1) To cut labels with a die. (2) The line of severence between a pressure-sensitive label and its matrix or adjoining label made by the cutting edge of a die.

Dispenser—A device that feeds pressure-sensitive labels, either manually or automatically, in convenient units. It often serves as a package for the labels as well.

Face material—Base Material. Body Stock. Face Stock. Any paper, film, fabric, laminated or foil material suitable for converting into pressure-sensitive label stock.

Glassine—A supercalendered smooth, dense, transparent or semitransparent paper manufactured primarily from chemical wood pulps, that have been beaten to secure a high degree of hydration of the stock. (Dictionary of Paper.)

Label—A slip of any material that can be affixed to anything, and indicating ownership, contents, directions, destination, or rating.

Label, face-cut—A die-cut label product from which the matrix has not been removed.

Label, laid-on—Die-Cut Label. Die-cut, pressure-sensitive labels mounted on a release liner from which the matrix has been removed.

Label, pressure-sensitive—Self-Adhesive Label. A pressure-sensitive (self-adhesive) label product is a die-cut part that has been converted through roll-fed production equipment utilizing the type of pressure-sensitive (self-adhesive) material that has a protective backing. The end product is produced in the form of either rolls, sheets, fan-fold, or by other techniques that produce like products that have been slit or cut from the converted roll. (Definition adopted at TLMI-Label Division meeting, January 30, 1964.)

Label, roll—Pressure-sensitive labels packaged in a continuous roll form.

Label, sheet—Pressure-sensitive labels packaged in sheets.

Label, tamperproof—Destructible Label. A class of pressure-sensitive labels that cannot be removed from a substrate without damaging them in some manner, thus making their reuse impossible.

Latex paper—Impregnated Paper. Saturated Paper. Paper manufactured by two major processes in one of which latex is incorporated with the fibers in the beater prior to formation of the sheet and in the second of which a preformed web of absorbent fiber is saturated with properly compounded latex. The papers are characterized by strength, folding endurance, resistance to penetration by water, flexibility, durability and resistance to abrasion. (*Dictionary of Paper.*)

Liner, release—Backing. Liner. Lining. Release Lining. The component of the pressure-sensitive label stock that functions as a carrier for the pressure-sensitive label. Prior to application, it protects the adhesive, and readily separates from the label immediately before the label is applied to its substrate.

Matrix—Ladder. Skeleton. Waste. The pressure-sensitive face material surrounding a pressure-sensitive label that is usually removed after die-cutting, thereby leaving laid-on labels on the release liner.

Metallized film—A plastic or resinous film that has been coated on one side with a very thin layer of metal by vacuum metallizing.

Pattern coated—Dry Lap. Separator. Strip-Coated. Zone-Coated. Refers to the width and spacing arrangement of strips of adhesive laid down parallel to machine direction and across the width of pressure sensitive label stock during its manufacture. Often directed to the adhesive pattern carried by pressure sensitive labels made from such material.

Release coat—Release Lacquer. The release liner treatment material that allows pressure-sensitive labels to release from the release liner.

Self-adhesive products—Converted pressure-sensitive labels and products protected by a release liner. Term used to differentiate from self-wound products such as pressure-sensitive tapes.

Shelf life—Storage Life. The period of time during which a product can be stored under specified conditions and still remain suitable for use.

Split back—Back Split. Slip Back. Split Liner. Slits in the release liner to facilitate its removal by hand.

Transfer tape—A pressure-sensitive adhesive (unsupported or reinforced) applied to a two-side release coated liner.

HISTORY OF THE INDUSTRY

The beginnings of the pressure-sensitive label industry are somewhat beclouded but it is generally conceded that it was in the period 1933-1934. Several companies were started at that time—one by Louis Fox; one by a pharmacist, Arthur Bennet (Kleen-Stik Products, Inc.); and a third by R. Stanton Avery (Avery Label Co.). Coincidentally, the birthplace of all three companies was California. In 1966 both

Avery and Kleen-Stik were still operating, and growing each year.

Shortly thereafter, two other companies appeared on the scene —Poster Products with a product called *Post-On*, and Chicago Show Printing Co. (*Mystik*). Both were Chicago-based and both operated under license of patents held by Ashley Fulton.

Three distinct marketing approaches were taken by these companies. Avery and Fox pointed toward the price-marking industry. *Mystik* and *Post-On* toward the point-of-sale and oil-change ticket field, and *Kleen-Stik* toward the field of transfer tapes for use on preprinted point-of-sale signs and labels. Prior to World War II these five companies comprised almost the entire industry and total sales probably did not exceed $1 million.

World War II almost put this infant industry out of business because of the shortage of both natural and synthetic rubbers and their strict rationing. Not until 1945 did the industry really begin to grow and to take its rightful and deserved place in the American economy.

WHAT *REALLY* IS A PRESSURE SENSITIVE PRODUCT?

Obviously, the definition of *self-adhesive products* as it appears in the preceding glossary—

Converted pressure-sensitive labels and products protected by a release liner. Term used to differentiate from self-wound products such as pressure-sensitive tapes.

as well as the *label, pressure sensitive*—

A pressure-sensitive (self-adhesive) label product is a die-cut part that has been converted through roll fed production equipment utilizing the type of pressure-sensitive (self-adhesive) material that has a protective backing. The end product is produced in the form of either rolls, sheets, fan-fold, or by other techniques that produce like products that have been slit or cut from the converted roll.

tell only a small part of the story. Let us analyze the three layers of a pressure-sensitive product individually.

First—"A substrate suitable for printing. . ." The most commonly used substrates are cast-coated papers, litho-coated papers, latex-impregnated papers, paper backed and bare metal foils, various plastic films such as cellulose acetate, "Vinyl," poly-ester films in clear, metallized or colors, certain textiles but mainly rayon satin, tag, bristol and vellum papers.

Most pressure-sensitive adhesives have a tendency to penetrate into paper and textile substrates; therefore the capable manufacturer of high quality pressure-sensitive materials must precoat or tandem coat these substrates with a "barrier" or "anchor-barrier" coating. A number of coatings suitable for this purpose can be commercially supplied by many manufacturers of liquid coating materials. Practically all plastic substrates as well as bare metal foil require the use of an anchor coat or a tie-coat, to provide a satisfactory bond with the adhesive. These coatings can be applied by any of several coating methods—viz. roller, reverse roll, mayer rod and even "knife over roll," depending on the particular chemicals and solvents being used. Normally a very light coating of a barrier coat suffices but a heavier lay down is usually necessary for the anchor coat.

Some manufacturers of pressure-sensitive materials buy their paper preanchor- or barrier-coated from the mills; some do their own coating in a separate or preparatory operation. In both cases a nonblocking type of coating must be used such as carbon methyl cellulose or zein. However, if these coatings are applied in a tandem operation with the adhesive coating and laminating following on immediately, a whole series of chemicals—even those coatings that are semi-pressure-sensitive in nature—may be used. This type of precoat has generally been found to be more efficient and also less expensive by the elimination of a preparatory prior machine operation. Quite obviously the coating material must be one that will not impair the "printability" of the substrate itself.

Second—the pressure-sensitive adhesive as defined earlier is also quite a complex chemical compound. Normally it consists of a natural or synthetic rubber, a "resin" or tackifying agent, a "plasticizer" or lubricating agent to keep both the rubber and resin ever-tacky and a solvent or mixing vehicle, which makes everything soluble and thus coatable and which is driven off in the drying oven. The most commonly used rubbers are "natural or smoked sheet," a polyisobutylene, GRS's of one or more types, acrylics and certain other specialized monomers or polymers. The most commonly used solvents are of the naphtha type, therefore highly inflammable and explosive and needing great care and safety measures in proper handling, the use of explosion-proof motors being mandatory in all mixing and coating equipment.

These adhesives normally are mixed in large churns; sometimes the rubber is milled with resins and plasticizers added to the mill batch; sometimes these are added in the churn at varying points in the mixing cycle. No common method is employed, each manufacturer doing it slightly differently.

Some types of pressure-sensitive adhesives lend themselves to the solid or calender method of coating, thus eliminating the need for a solvent and drying ovens. Most of the present day surgical tapes, bandages, and plasters are produced by this method as well as many industrial tapes. However, this method is not generally used by the makers of pressure-sensitive label papers.

The formulation of any solvent-type adhesive must be done with a careful view to the drying cycle. For example, it is quite easy to over-dry a paper substrate, taking out all its moisture, and resulting in a brittle, curly sheet almost impossible to handle on any printing press, be it roll- or sheet-fed. It is important to remember that paper is normally 5 to 6% water.

Third—release paper. Historically, the first release liner used was a "rubber holland cloth," a cotton cloth heavily impregnated with starch and highly super-calendered. The starch itself acted as the releasing agent from the adhesive. For quite a few years this was universally employed but its high cost held back more widespread use. Furthermore, its releasing characteristics were a limitation to the use of higher tack adhesives. Thus it was apparent that a great need existed for a lower cost and a more efficient release liner. The first step away from holland cloth was to a vegetable parchment paper to which the pressure sensitive manufacturer would apply a release coating. The early release coatings varied between the manufacturers, but were primarily either dextrines, lacquers or waxes, or a blend. This combination of parchment paper and release coating primarily accomplished the desired cost reduction but did not yet bring about the efficiency of the release factor that was so urgently desired by the industry. For some years, primarily because of the release factors, only a relatively low tack or removable type of adhesive could be used. However, it was generally known that a permanent or tamper-proof type of adhesive would greatly expand the market for pressure-sensitive products. Formulation and coating of this type of high-tack adhesive was fairly easily accomplished by all producers, but it was not until the advent of Silicone that a really efficient release coating was developed.

Strangely, prior to Silicone the efficient manufacture of a pressure-sensitive paper was an art rather than a science. The very specialized knowledge and the secrecy of the Release Coating was the really determining factor in measuring the relative quality of the various products on the market. Silicone changed all this and today what was an art has become common knowledge. Just as there are quite a number of manufacturers of pressure-sensitive papers, films and foils, so there are quite a few manufacturers of release papers.

APPLICATIONS

The applications for pressure-sensitive products can be divided roughly into three main categories:

Labels —	Advertising
	Identification
	Specification
	Data-processing
	Mailing
	Price-marking
	Closures
	Reminder
	Fill folder
Signs —	Advertising
	Window
	Truck
	Indoor
	Outdoor
	Warning
	Decals
	Oil-change reminders
	Shelf-talkers
Forms —	Order
	Work-order
	Reporting
	Envelope
	Data-processing
	Hospital

The sale of finished, converted pressure-sensitive material to the end user today somewhat resembles the container industry in that there are a number of "vertical" as well as "horizontal" companies. A vertical company is one that produces the entire pressure-sensitive product, making everything except its own papers, and prints, die-cuts, and sells the finished label or sign directly to the end-user.

A horizontal company is one that manufactures only the pressure-sensitive materials themselves, delivering these to printers for printing and die-cutting, for sale *by the printer* to his customer, thus functioning solely as a supplier to the printer either by direct sale to the printer or through established and usually franchised paper merchant houses.

Just as in the container field, there are also those that "carry water on both shoulders," so to speak, and sell to the printer when possible, but also directly to the user when necessary.

Thus quite a few choices are available to those that contemplate entering the pressure-sensitive paper business, the selection largely dependent on the type of marketing organization at their disposal.

In addition, a printer or paper merchant actively selling pressure-sensitive products has a choice of suppliers who either compete with him at the end-user level or function solely as a raw-material supplier.

DEVELOPMENTS IN DISPENSING MACHINERY

The tremendous growth of pressure-sensitive labeling has really only occurred since 1959. Its spring board has been the development of

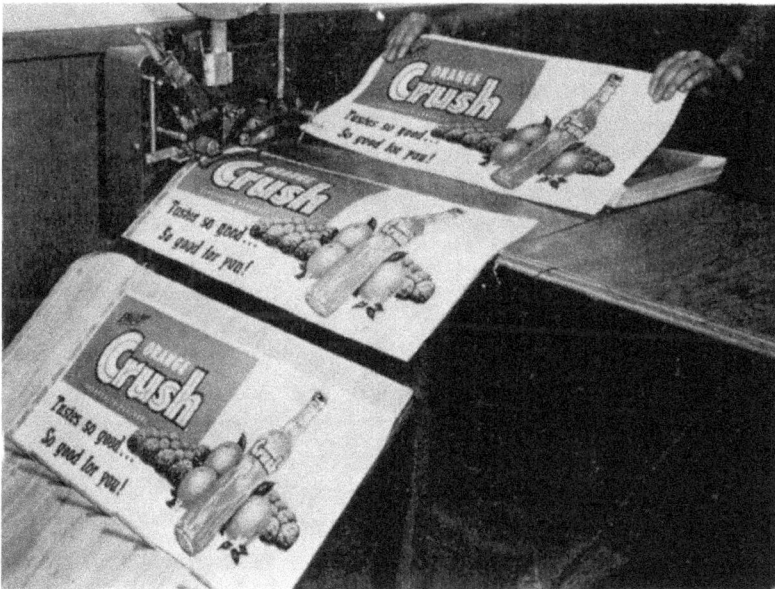

Fig. 11-1. Automatic tape applicator applies strips of pressure-sensitive transfer tape to signs and banners. When ready for use, backing is peeled away to expose adhesive and sign is pressed into place on window or wall (*Courtesy of KLEEN-STIK PRODUCTS, Inc.*).

very efficient automatic pressure-sensitive labeling equipment. Today standard machines are available that have a capability of applying pressure-sensitive labels at speeds up to 800/min—actually far exceeding the speeds of equipment that employs other adhesive-coated labels.

Obviously automation has thus reduced the price of a pressure-sensitive label—*on the article*—to a level where it is actually the least expensive as well as the most efficient method of labeling. A recent cost study was made by one of America's large food processing companies. It was paying $1.60-$1.80 per thousand for heat seal labels. The time cycle to heat and apply the label limited the labeling line to a speed of 100/min, whereas speeds of more than 300/min. were available with a pressure-sensitive label applicator. Thus label buyers, industrial engineers, and process engineers must analyse the *cost* of the label *after application* versus the cost of the label itself.

Too many graphic arts people are too price conscious about the wrong things. A really good label salesman carefully analyzes the various needs and requirements of his customer, the method of label application, the material the label must stick to, the end use of the label

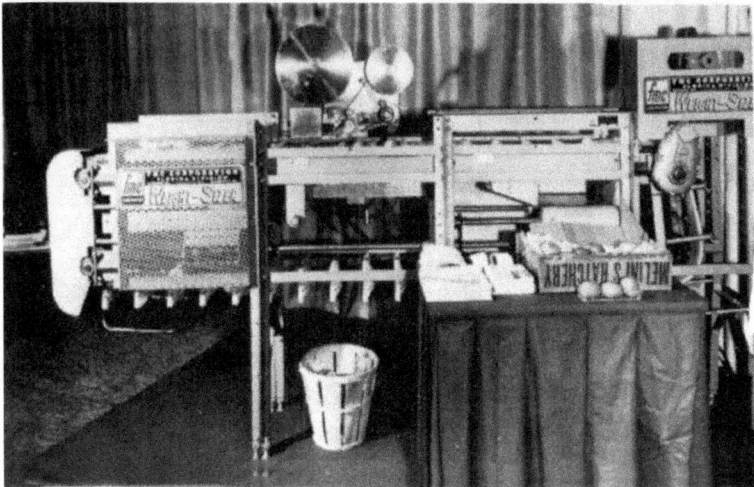

Fig. 11-2. Compact Label-Aire® applicator installs easily on many standard packaging units such as this FMC Weight-Sizer (*Courtesy of KLEEN-STIK PRODUCTS, Inc.*).

Fig. 11-3.   Basic semiautomatic labeling system (*Courtesy of Avery Label Company*).

Fig. 11-4.   Automatic labeler for book distributor (*Courtesy of Avery Label Company*).

Fig. 11-5. New Avery Model 300 labeler (*Courtesy of Avery Label Company*).

whether it should be permanent or removable, how much abrasion it must withstand, if any, and extremes of temperature and humidity. Once all these features are determined it is quite surprising how often a pressure-sensitive label is the only efficient type to use, and its cost *on the product* is frequently the lowest of the costs of various available types of labels.

Point-of-Sale Signs

Pressure-sensitive signs cover a wide area of applications and uses—from the small ($1\frac{1}{2}$ in. $\times 3$ in.) oil change sticker applied to the door-jamb of an automobile to the large identification signs (5 ft $\times$ 8-12 ft) to replace painting the truck itself; from "Shelf Talkers" (1 in. $\times 7$ in.) to window and wall advertising posters in the retail stores.

A wide range of pressure-sensitive materials is available from practically all manufacturers, from cast-coated papers, to polyester and PVC films and from low-tack or removable adhesives to high tack (permanent) type adhesives. Thus the proper substrate and adhesive for the particular end-use are readily available to the printer, and usually through a local paper merchant from stock in most instances.

The growth of pressure-sensitive materials since 1956 is such that stocking several grades and in several different sizes has become mandatory for any up-to-date paper merchant or silk screen supply jobber.

Transfer Tapes

These are made by a number of manufacturers today and are readily available in many widths from local paper merchant stocks as well as on special order.

What are Transfer Tapes?

A transfer tape is a roll of dry adhesive wound on a two-side release coated paper. It is manufactured primarily in two adhesive grades —"Low-Tack" and "Hi-Tack." Normally the "Hi-Tack" product is one where the adhesive layer itself is supported by a web of tissue and interleaved with the release liner. The "Low-Tack" normally has no web, the adhesive itself being clear and water white, but also interleaved with release liner.

Both are available in widths of ½ in. to 54 in. and in lengths from 60 to 1000 yd, depending on the manufacturer and adhesive grade wanted.

How can "Transfer Tapes" be Applied?

There are three basic methods of application:

By Hand—Obviously the first and most costly method.

Roll to Roll—By merely spotting the tape on a roll holder that can easily be attached to any roll-to-roll or roll-to-sheet piece of equipment; the tape or tapes are merely allowed to flow onto the web at the desired location. Any number of rolls can be mounted thus so that the desired number of strips of tape can be applied either on a printing press, sheeter, or even collator.

This method provides *only* continuous strips of tape.

Sheet Application—Machines are available for the application of transfer tapes to sheeted material. These machines are available in many trade service plants throughout this country and abroad and are all listed in local telephone books. Machines less complex are also available for sale or lease directly to printers that have sufficient volume of work to warrant the investment. Prices on these machines run from $1000 to $10,000 depending on width.

PRE-DIE-CUT SHEETS

Several manufacturers of pressure-sensitive label stocks are experiencing quite a rapid growth in the sale of sheets of pre-Die-cut labels or press-ready sheeted labels. They are available in a number of sheet sizes and a number of label sizes. Because they are pre-die-cut the printer has a comparatively simple plate lay out and register problem.

Further advantages are the elimination of a secondary operation and better sheet feed through the press and packaging right off the press because there is no adhesive at any of the four sides.

These press-ready sheets are ideal for use on most office duplicating machines and in other plant and office equipment.

One common method of producing the sheets is on a roll-to-sheet blanking press, which die-cuts and removes the skeleton from around the labels and sheets all in one operation and most probably from either steel rule or engraved dies.

Press-ready sheets offer a fine opportunity for a specialized and profitable product line.

SHEETS WITHOUT SPLITS

Probably the newest development in the pressure sensitive paper field is that of two new types of release paper construction, both of which eliminate the need for either "splits" or "die-cutting" a tab.

One construction involves the use of a silicone-coated, machine-creped, polyethylene-backed kraft, which can be removed by merely squeezing between the thumb and forefinger to break the bond between the adhesive and the release paper.

Another development is that of an overall pattern of acid ink printing that eats partially through the release paper. When bent along any one of these printed lines the release paper splits, thus providing a start for peeling it away.

Both of these inventions were pointed toward the elimination of a printer's having (1) to lay out his printing to conform to the split patterns available, and (2) separate die-cutting of a peel tab when ordinary guillotine cutting would suffice. These new constructions will no doubt increase still more the use of "pressure-sensitive products."

Release Papers

WILLIAM LOUDEN

Many raw materials and articles of commerce are either permanently tacky or have some tendency to stick to other surfaces at some stage of their manufacture or storage. When this happens, it is usually necessary to cover the surface of the product at least temporarily until the tackiness is gone or until the product is ready to be used. For these purposes the so-called "releasing papers" are almost always employed. A release paper may therefore be defined as a web, composed at least partially of cellulose fibers, that shows low enough adhesion to some other material so that it may be removed easily without damage to either the paper or the product.

A surprising number of products have the ability to stick to other surfaces. Unless some precaution is taken, all frozen meats and many frozen pastries and vegetables will stick to the container. Candies, pastries and breads, candied fruits and many cereals will also stick to many surfaces. In the industrial world, almost all rubber and plastics, all adhesives, natural and modified wood rosins, asphalts, concrete and paints, have the ability to adhere to other surfaces. In most of the above-named cases, there are many separate uses for releasing papers, each with its own list of performance requirements.

As our manufacturing and distribution techniques become more complex, the needs for releasing papers multiply at a rapid rate.

One large manufacturer of chemicals used to produce releasing papers estimates that the amount of releasing papers treated with his class of product alone exceeded 150 million lb in 1964.

TYPES OF RELEASE PAPERS

Three general classes of release papers or boards are considered:

1. Those that release by virtue of a smooth, impenetrable surface only.
2. Those that release by transferring a layer of coating or treatment to the tacky surface.
3. Papers that are chemically treated to reduce adhesion but that do not contaminate the product.

Ordinary paper has almost no valuable traits for releasing purposes.

It is a mat of fibers with low resistance to penetration by most liquids. If a liquid (such as water, organic solvents, melted wax, or an adhesive) contacts the paper, penetration occurs almost immediately. If the material later becomes solid, as in freezing or drying, a very strong adhesive bond is formed, sufficient to prevent removal. In many cases, the simple application of a smooth, dense impenetrable film or coating to the surface of the paper provides the necessary barrier characteristics. This situation is particularly true with frozen meats and vegetables. Waxed or polyethylene coated folding cartons provide sufficient release in most instances. In the case of waxed cartons some tendency toward flaking occurs.

Good release is an essential property of locker papers, and in all instances release is accomplished by means of an impenetrable film lamination, or a coating, with no additional treatment necessary.

Most plastic films have fairly good releasing properties in food applications, by virtue of their smoothness and impenetrability. Release coatings are rarely needed for such films.

Laminations of aluminum foil to paper have been used to effect release by the same mechanism. Plastic films such as cellophane and aluminum foil laminates have also been widely used in the plastics industry as separator sheets in the manufacture of polyester-based and melamine-based laminates. These materials also provide specific surface finish effects, which are desirable. Sometimes release is marginal, forcing application of release treatments as well.

Vegetable parchment is a unique type of paper that has a number of advantages for releasing applications. An absorbent paper is first manufactured, similar to blotting paper or toweling, and then dipped in 68% sulfuric acid that has been held at a temperature of 50-60°F for a few seconds. During this short period, a portion of the cellulose is solubilized to a gelatinous state, which fills the capillaries between the remaining fibers and covers the surface. If the paper were left in the acid long enough it would be completely solubilized. In the normal process, the paper is exposed for a few seconds and then dipped into 35-37% acid, which stops the action. The paper is washed and dried. It has a fiber-free surface and high resistance to penetration by nonaqueous organic materials. Without further release treating it can be used as a separator sheet in the manufacture of polyester laminates and also in the manufacture of some types of rubber belting.

Sodium silicate has been used as a coating on standard unbleached kraft paper to bond the fibers together to prevent picking and to provide a semi-impervious surface. These types of papers are not subjected to any release treatments and are generally effective only in mild releasing applications, such as interleavers for compounded sheet rubber and rubberized fabric that are later die-cut into shoe soles and other shapes in the manufacture of shoes and overshoes.

Release papers that rely on the principle of gross contamination to effect separation do not find much use. However, a special clay coating was developed some years ago. It was very high in clay and contained just enough binder to adhere the coating to the paper. Coating weights of 15-30 lb/3000 ft^2 of surface area were common. This coating was applied to a standard grade of bag kraft and used as the inner liner of a multi-ply bag for packaging various types of synthetic rubber. When the paper was removed from the block of rubber, a layer of clay was left behind. Because the rubber was later compounded on a rubber mill, the presence of the clay was not objectionable. This same type of paper is sometimes laminated to the inner wall of spiral-wound containers used to package roofing asphalt and certain natural rosins. In most applications, the contamination is objectionable.

Waxed papers are sometimes used as release papers in certain types of printed decalcomanias. A clear layer of varnish is first laid down, followed by printing ink and finally a heat-activated adhesive system. In the manufacture of china, printed designs are transferred to the hot, unfired china. The heat activated adhesive adheres to the china and the wax melts. The decalcomania literally "floats off" the release paper. As the china is later fired, the wax contamination is not too serious. This same principle is used in other types of decalcomanias where low cost is important and the wax contamination is not objectionable.

Chemically treated release papers are the most widely used.

CHEMISTRY OF RELEASE TREATMENTS

Silicones

By far the most potent class of release treatments is the silicones. If applied in the prescribed manner, these materials will reduce the forces of adhesion by 98 to 100%, depending upon conditions. Too much reduction in adhesion can be undesirable in many cases. Hence the search for release agents to provide different degrees of adhesion continues. Other release treatments have been developed and will

be outlined later. Silicones are by far the most widely used release
agents. Their biggest drawbacks are high cost and relatively low cure
rates, necessitating cures in the range of 350 to 450°F for 8 to 15 seconds, depending on conditions.

De Monterey has presented an excellent summary of the chemistry
of silicone release treatments, giving a brief insight into the theory of
why they work so well.[1]   A number of patents have also been issued.[2-3-4-5-6-7]

As known for at least twenty years, dimethylpolysiloxane polymers,
if applied in a thin film on most surfaces, will drastically reduce adhesion. They are composed mainly of repeating units of the following
structure:

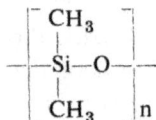

$$\left[\begin{array}{c} CH_3 \\ | \\ -Si-O- \\ | \\ CH_3 \end{array} \right]_n$$

Dimethylpolysiloxane polymers are inert and have outstanding heat
stability. Methyl groups attached to the silicon atom are believed to
have a high degree of rotational freedom, far more than their stereoically
hundred carbon analogs ($C-CH_3$). The polymer is highly ordered
within each polysiloxane molecule, which is considered to be helical
or spheroidal in structure. The high degree of orientation of the
relatively inert methyl groups, combined with the extreme stability of
the siloxane groups, probably explains why dimethylpolysiloxanes
are so effective in reducing adhesion.

Unfortunately, dimethylpolysiloxanes show very little reactivity
toward cellulose. When applied to even dense papers, such as vegetable parchment, they penetrate the surface. They also migrate to the
tacky surface and produce undesirable effects. The release action can
be temporary or greatly reduced if too much silicone leaves the surface
of the paper. To function effectively, the silicone must be applied to
the surface and held there by curing or some other mechanism to
render it nonmigratory.

The first really successful silicones used for release paper treatments
were the methyl hydrogen polysiloxanes:

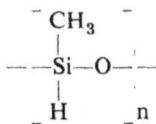

$$\left[\begin{array}{c} CH_3 \\ | \\ -Si-O- \\ | \\ H \end{array} \right]_n$$

Solutions of these materials in toluene, xylene, and hexane were mixed with small percentages of zinc octoate or stearate or stannous octoate or oleate catalysts. Under the influence of relatively high temperatures (350 to 450°F) these systems would cross-link and adhere to the paper surface, so that the overwhelming percentage of the treatment would remain fixed on the surface, inert and nonmigratory. Unfortunately, the curing rates of these systems were extremely low, not at all adapted to commonly accepted coating speeds and temperatures in the paper industry. The cure would continue at room temperature for several months, and would eventually lead to a slight loss in release efficiency, under certain conditions. As a consequence, substantial quantities of inadequately cured silicone-treated papers were produced. These showed both contamination and loss of release efficiency, particularly after the paper had been in contact with the adhesive for long periods of time. Clearly, further improvements were needed.

The real breakthrough came when polysiloxane fluids end-blocked with hydroxyl units were blended with methyl hydrogen polysiloxanes. In the presence of suitable catalyst and sufficient temperature, the methyl hydrogen polysiloxane crosslinks the fluid and also reacts with the cellulose to form a good bond.

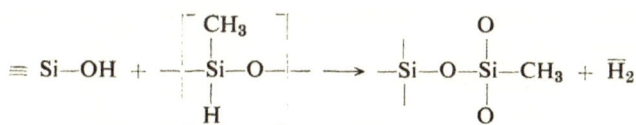

$$\equiv \text{Si}-\text{OH} + \begin{bmatrix} \text{CH}_3 \\ | \\ -\text{Si}-\text{O}- \\ | \\ \text{H} \end{bmatrix} \longrightarrow \begin{matrix} \text{O} \\ | \\ -\text{Si}-\text{O}-\text{Si}-\text{CH}_3 \\ | \\ \text{O} \end{matrix} + \overline{\text{H}}_2$$

This new type of system was much faster curing and when cured to a rubbery polymer was less prone to contamination. It is much more adaptable to normal paper coating speeds and gives much more reproducible results. Most commercially available silicones recommended for release coating are based upon this general composition. A slight amount of release efficiency has been sacrificed.

Silicone fluids of higher viscosity are more desirable for releasing purposes, both from the standpoint of efficiency and lack of contamination. Linear polymers used for release purposes vary in molecular weight from 5000 to 250,000. A "stripping" technique has been devised where the lower molecular weight fractions can be removed. This procedure gives a superior product for release purposes, particularly from the standpoint of contamination.

Types of Commercially Available Silicones

Silicones may be applied in emulsified form in operations where the presence of solvent is objectionable. The silicone and the catalyst are emulsified separately, generally by the chemical manufacturer. The emulsion systems are fairly stable under the influence of mechanical agitation. The cost of application is generally lower than for the solvent systems. Emulsions of methyl hydrogen polysiloxane and also the faster-curing later variety described previously are commercially available. The methyl hydrogen polysiloxane system is considerably cheaper. When producing releasing papers that will contact food, a special catalyst emulsion system is required. Both types of silicones are also available as 100% solid materials so that they may either be emulsified or dissolved in special solvent systems.

The newer type of faster-curing silicone is also available in solvent solutions, generally xylene, toluene or perchloroethylene. Hexane, toluene or perchloroethylene are used for additional dilution. In the solvent systems, a more complete mixing of silicone and catalyst is assured, and the solution is deposited on the substrate as a continuous film rather than as a layer of discreet particles that later blend togegher. Release papers with higher gloss coatings can be produced by means of the solvent technique. Although operating costs are higher, more uniform release papers can be made, and adhesion to the substrate is said to be better.

Silicones as normally prepared have a fogging effect on photographic film, which can be objectionable. Special grades that do not suffer from this disadvantage are available in either solvent of emulsion form.

In some applications, the silicones deliver too much release. After considerable research, special additives have been developed that can be blended with standard silicones to reduce their release efficiency. The composition of these additives has not been disclosed but they are generally believed to be modified polysiloxane polymers. It is possible to change release rather markedly with these agents, but minor adjustments are not practical because the curve is too steep.

Special silicones have also been offered where every effort has been made to reduce adhesion to as near zero as possible. The problem is to secure maximum release without sacrificing adhesion to the substrate.

Special adhesion promoters have been developed to promote a good bond between the silicone polymer and cellulose. Also silicones that cure by moisture and heat and that do not require a catalyst are available.

On the whole, silicones are effective release agents for most tacky materials, except silicone rubbers, sealants, and adhesives.

Chromium Complexes

In 1941, Ralph K. Iler was granted U. S. Patent 2,273,040[8] covering a unique new class of chromium complexes of particular interest to the paper industry. World War II interrupted commercial development, but in 1949 stearato chromic chloride was offered by Du Pont under the trade name Quilon S®. The following structural formula has been advanced:

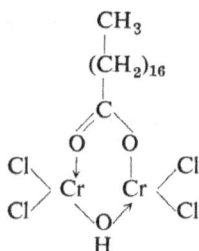

$$CH_3$$
$$(CH_2)_{16}$$
$$C$$

The complex is offered as an isopropanol solution but it can be diluted with water. Some of the chloride groups hydrolize and are replaced by hydroxyls. The chromium end of the molecule has a strong affinity for cellulose and other polar surfaces. On parchment, for example, a portion of the complex is irreversibly adsorbed. On curing for a few seconds at temperatures of 180 to 220°F, hydrochloric acid is liberated, and the complex loses its solubility in water. Urea, ammonia, hexamethylene tetramine or mixtures of urea with sodium formate and formic acid are used with the aqueous solutions of stearato chromic chloride to act as neutralizers or acceptors for the acid and to prevent embrittlement of the paper.

Once the cure is affected, paper treated with stearato chromic chloride shows pronounced water shedding properties and, depending on the density of the paper itself, releasing qualities. The most pronounced releasing effects are observed on dense papers such as vegetable

® Registered trade mark, E. I. duPont de Nemours & Co., here and on following page.

parchment or greaseproof paper. For these types, adhesion of standard pressure sensitives can be reduced to roughly 40 to 60% of thea dhesion measured on untreated samples.

Many believe that the complex orients on the surface of the paper, wherein the nonpolar methyl-terminated "tail" tends to face the adhesive surface, and the reactive chromium is bonded to the cellulose.

Myristato chromic chloride is also commercialiy available under the trade name Quilon M®. Its properties are similar to those of stearato chromic chloride. A special complex with lower acid content that works well with an ammonia neutralizer system has been offered under the designation Quilon C®.

The chromium complexes offered commercially may be considered as intermediate release agents. They are considerably less expensive and much easier to apply than the silicones and rarely affect the tack of adhesive systems. They provide fairly effective release action from silicone rubbers and silicone-based pressure sensitives.

Releasing papers treated with chromium complexes are quite effective against many plastics, such as polyesters and melamine-formaldehyde-and phenol-formaldehyde-based laminates. They are also in food applications, particularly as bakery tray liners.

Alkyl Branched Polymers

Certain polymers with long-chain alkyl branching show releasing characteristics when applied as a continuous film. Polyvinyl esters of higher fatty acids containing sixteen carbons or more are effective. Copolymers of vinyl acetate or vinyl alcohol and vinyl stearate are an example. The molar percentage of vinyl stearate can be as low as 30, but the preferred value is not less than 60%.

A polyvinyl ester of a higher fatty acid, such as stearic, differs markedly from a polyvinyl ester of a lower acid such as acetic. The polymer esters of the higher fatty acids have sharp melting points, as compared with polyvinyl acetate, which softens and gradually melts over a relatively wide temperature range. Many believe that the higher polyvinyl esters form crystal micelles in which the long side chains are packed together in parallel alignment. When the polymer melts, these micelles are probably disrupted. Films of this type show release properties below the melting point only.

Alkyl-branched polymers are classed as intermediate release agents. Their main use is as coatings for the back side of pressure-sensitive tapes.[9,10,11,12,13,14,15]

Alkyl-Substituted Amines, Amine Salts, and Quaternaries

These materials can produce some releasing action. They tend to be adsorbed on the paper surface and to orient, probably by a mechanism similar to that for the chromium complexes. They may be regarded as intermediate release agents. Webber[16] has found that the octadecyl amine salt of mono-ocatadecyl acid orthophosphate and the octadecyl amine salt of mono-dodecyl acid orthophosphate are effective. Main use is for backside coatings on pressure sensitive tapes.

Proprietary Release Coatings

A number of coatings formulators have produced blends of various kinds of release agents and film formers. Copolymers of certain silicones and alkyd resins have been made. In general, these are offered where intermediate release levels are needed or where special effects, such as high gloss, are desired.

PAPER PROPERTIES THAT INFLUENCE RELEASE

Although the effectiveness of the release treatment has much to do with the success or failure of a release paper in a particular application, the exact characteristics of the paper itself are equally important. Paper as normally made is a porous mat of cellulose fibers that can be readily penetrated by the release treatment or the tacky material itself under conditions of actual use. In many plastic applications, the paper can become thoroughly impregnated if it does not possess sufficient barrier resistance.

The so-called "dense papers," such as vegetable parchment, glassine, and greaseproof, are very well suited for release treatment. When applied from either aqueous or solvent solution, the release treatment is almost entirely confined to the surface of the paper, where it is needed. In most cases, less release treatment is needed for optimum results. These papers also yield release mediums that are uniform in release qualities from one area to another or from production run to production run. Less dense papers have a tendency to adsorb the treatment in an irregular manner that can lead to irregularity in release.

Techniques have been developed for the application of release treatments to more porous types of papers. Where aqueous systems are involved, such as solutions or emulsions, high viscosity film formers, such as polyvinyl alcohol, sodium carboxymethyl cellulose or carboxymethyl starch are mixed with the release treatment itself. The resultant

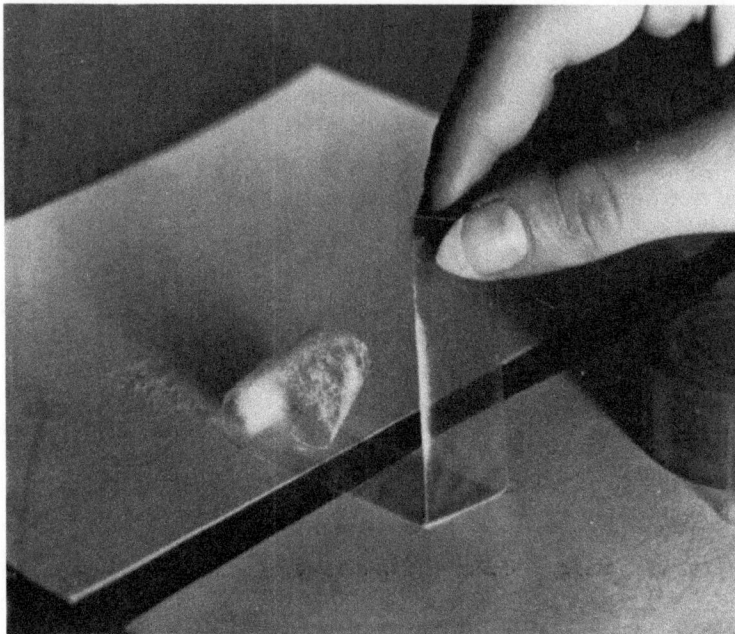

Fig. 11-6. Fiber picking is eliminated by application of a silicone/carboxymethyl cellulose blend to the surface of ordinary paper (*Courtesy Dow-Corning Corporation*).

high viscosity coatings show reduced tendency to penetrate the paper. The film former also tends to bond the fibers together and prevent fiber picking when the tacky material is removed. Such coatings are applied by air knife coater for best results. (See Fig. 11-6.)

Sometimes the film former is applied to the paper first and dried, followed by the release treatment. A continuous coating gives the best results. However, the film former can be applied at the size press of the papermaking machine, followed by drying and calendering or super-calendering. The net effect of this procedure is to provide a barrier to penetration of the release treatment and to bond the fibers together. Care must be taken to make sure that sufficient film former is applied uniformly on a sustained basis.

When solvent-based coatings are applied, the substrate must have sufficient holdout properties to keep the coating on the surface. Some form of pretreatment, either sizing or coating, is necessary before the application of release treatment. A special grade of kraft paper impregnated with sodium carboxymethyl cellulose and calendered is widely used. Polyethylene coated papers are also used for release coating. If polyethylene is applied to both sides, papers of exceptional dimensional stability are produced because moisture cannot penetrate the paper and cause the fibers to expand. Sometimes kraft papers are supercalendered and then varnished with thermosetting materials that are inert towards the adhesive surface.

In all cases where special coatings or sizes are applied, the materials used must be incompatible with the tacky product, or have a low order of compatibility and should not react with the adhesive over a fairly broad temperature range.

The ideal substrate for solvent or aqueous release coating should be tough and strong, flexible, and should retain these properties at low relative humidities. It should have good dimensional stability and should possess a smooth impervious surface that has low solubility or compatibility with most organic materials and is not reactive. This surface should be readily wet by the release treating solution but should be fiber-free.

TESTS FOR RELEASE QUALITY

When used for purposes of research or product development, a release test should throw some light on the usefulness or efficiency of a particular release paper for some hoped-for application. When used for quality control purposes, it should be able to spot unacceptable material. When the literally hundreds of different uses for release papers are considered, it is easy to see why no one release test will ever gain universal acceptance.

When controlling quality for some purpose, the best test is one that most closely approximates the conditions of actual use. Thus, in the manufacture of plastic laminates, an actual laminate must be made under controlled temperature, time, and pressure before a reasonable prediction can be made as to whether a particular release paper can be removed without difficulty. Similar conditions exist when release papers are used for bakery tray liners or interleavers for frozen hamburger patties.

There are a number of good testing procedures to determine the relative effectiveness of the release treatment, once applied to the paper, and the ability of the treatment to resist adhesion buildup under accelerated aging conditions. These same tests can check uniformity of release and the amount of contamination a release paper may cause.

The Keil Test

Test Equipment.

To carry out the test, the following materials and equipment are necessary:

1. A roll of adhesive tape 1 in. wide (Johnson and Johnson Red Cross waterproof adhesive tape). As the test results are comparative, all tapes should be cut from the same roll for any one set of tests.

2. A 4½-lb standard tape roller* to apply each tape uniformly and with equal pressure.

3. Metal plates and weights sufficient to produce a pressure oz ¼ lb/in.^{2} of the test tape. This condition simulates storage under pressure. Four aluminum plates, each 4×6 in. are supplied with the Keil Tester,** shown in Fig. 1-7. Also supplied are three weights, 24 oz. each, to produce the required pressure.

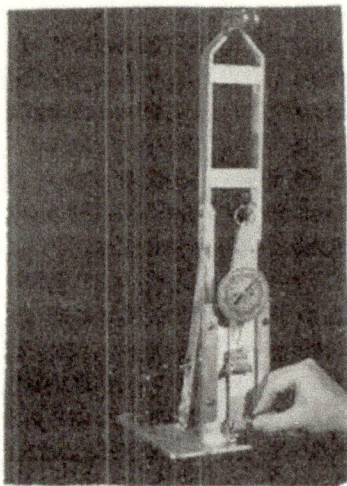

Fig. 11-7.   Keil Tester (*Courtesy of Dow-Corning Corporation*).

---

* Available from U.S. Testing Co., Hoboken, N.J.
** Available from Dow-Corning Corporation, Midland, Michigan.

4. An oven capable of maintaining a temperature of 70°C (158°F) for 20 hours is used to age the test samples to a degree comparable to storage for one year at room temperature.

5. The Keil Tester Model 2, or similar machine. Essentially, the Keil Tester is designed to strip tapes from test surfaces at a constant speed pull of 12 in/min. The pull is measured on a dial spring balance as the tape is stripped. Two balances are packaged with each Keil tester: one with a capacity of 500 g, the other with a capacity of 2000 g. Also supplied with the Keil tester (locked in place at the base of the machine) is a stainless steel test panel 3 in. × 6 in. × $\frac{1}{16}$ in. This panel serves as the control surface to determine the "subsequent adhesion" of the tape—its adhesive qualities after contact with the coated paper.

Test Methods

1. **Measurement of Anti-Adhesiveness.** (a) Strip the first three layers of adhesive tape from the roll and discard. Next cut a minimum of three six-inch strips of adhesive tape from the roll and apply them to the release paper, with the adhesive side in contact with the paper. Cut the paper into strips slightly wider than the tape.

   (b) Place the tape-on-paper laminate between two flat metal plates, and add enough weights to produce a pressure of $\frac{1}{4}$ lb/in.² of tape.

   (c) Age the samples in an oven at 70°C (158°F) for 20 hours. Remove the samples from the oven and allow to cool.

   (d) Lift one end of the tape from the coated paper and fasten the tape to the balance. Then fasten the paper from which the tape was lifted to the fixed clamp at the base of the tester. See Fig. 11-7.

   (e) Start the Keil tester and observe the dial reading on the balance at 1-in. intervals as the tape is stripped from the paper. Average the five readings and record the average as release force in g/in. of width. Save the tapes for use in measuring subsequent adhesion (see next paragraph). This test should be carried out immediately after the tapes are stripped from the paper.

2. **Measurement of Subsequent Adhesion of Test Tapes** (a) Use a 3 × 6-in. stainless steel test panel. Clean it thoroughly with reagent-grade toluene and allow it to dry. Lift the panel

by the edges so that finger marks will not reduce the adhesion of the tape.

(b) Take one of the 6-in. tapes previously stripped from the paper and apply it to the stainless steel panel with two passes of the tape roller. Do not use extra pressure, but only the free weight of the roller.

(c) Insert the panel in the holder at the base of the test machine, pull up the bottom end of the test tape, and fasten this end to the balance.

(d) Strip the tape from the metal panel and record the dial readings as described in Section 1, step (e).

(e) Repeat steps (a), (b), (c), and (d), Section 2, for each test tape released from the coated paper, and average the results.

(f) As a control, repeat steps (a), (b), (c), and (d); use a 6-in. sample of tape taken directly from the roll.

(g) Compare the test results: force required to pull tapes that have contacted the release coating vs. force required to pull the tape that did not contact the release coating. The results should be approximately equal.

Pressure Test

The pressure test is similar in many respects to that developed by Keil. Johnson and Johnson adhesive tape is used. 1-in. wide strips of tape are applied to the release paper test specimen. Four or five samples are prepared. These are pressed in a Carver laboratory hydraulic press for 2 minutes at 400 lb/in.² and removed. Stripping force is immediately measured on a standard pendulum type tensile tester at 180 peelback and a stripping rate of 12 in./min. Results are reported in g/in. of width. Tests are conducted at 72°F.

In a second test, the samples are pressed at 400 lb/in.² as described, but instead of being tested immediately, they are hung in an oven for 7 days at 150°F and then cooled to room temperature and tested as just described. A thorough study has shown that 5 to 7 days are necessary for development of maximum adhesion.

When a fresh supply of adhesive tape is secured, adhesion to a clean stainless steel panel is always checked. Application at 400 lb/in.² followed by immediate stripping is recommended. Considerable variation in adhesion to steel is common, depending upon the particular batch of tape. Avoid tape with low adhesion values.

In these various pressure test procedures, adhesion is sometimes reported as grams of stripping force per inch of width. Sometimes it is expressed as a ratio (Steel Ratio):

$$\text{Steel Ratio} = \frac{\text{adhesion to paper, g}}{\text{adhesion to steel, g}}$$

Adhesive Casting Tests

For work with pressure-sensitive adhesives, an adhesive casting test procedure is highly recommended. First, it is possible to run tests on the actual adhesive to be used. Second, the test is more severe, particularly at high rates of stripping, and can pick up slight differences not otherwise discernible.

In this procedure, a solvent solution of the adhesive is actually coated onto the release paper. Sufficient adhesive should be applied to provide a thickness of dry adhesive film of 2 mils. Allow the coated sample to air-dry for 2 minutes at room temperature, followed by one minute in an oven at 200°F. Remove and apply Mylar or cellophane film to the adhesive using the 4¼-lb standard tape roller on 1-in. cut strips. Samples can be peeled back slightly and then tested in a standard tensile tester (Keil tester) or by other techniques. The sample strips can be hung in an oven for 7 days at 150°F before testing to judge roughly the effect of prolonged aging.

Stripping Rate

The force required to remove most adhesives from release papers usually depends upon the stripping rate. Generally speaking, the slower the paper is removed, the lower the force of adhesion. The most common stripping rate is 12 in./min but there are some test procedures that employ speeds as high as 1800 in./min (150 ft/min). High stripping rates are useful in picking out slight differences in release, particularly for aged samples. In many applications, the release paper is removed rapidly, as in printing, die cutting and removal of "lace" in the manufacture of pressure-sensitive labels. All these operations are accomplished in one pass through a flexographic press.

Stripping Time Tests

Sometimes the interval of time necessary to remove a given length of release paper under a given load is determined. The adhesive and release paper are brought together by means of any of the techniques previously described. The release paper is peeled back an inch or so and attached to a weight, usually 75 or 100 g. Peelback is often

180°. The length of time required to remove four inches is determined. Very good release would give values of 1 to 2 seconds with the 100-g weight. The stripping rate here would be 120 to 240 in./min. This test approximates actual application conditions in certain label dispensing machines.

Stain Tests

Most release treatments impart water-shedding qualities to paper, particularly the silicones and the chromium complexes. A special cellulose stain called *Texchrome* sold by Fisher Scientific Company, New York, N.Y. is widely used to check uniformity of treatment across a surface area. The release paper is smeared with dye and wiped off immediately with absorbent paper. The effectiveness of the coating can be judged by the amount of color retained by the test strip. With an effective coating the paper shows very little staining. Uncoated paper would be dyed a deep purple. Poorly treated areas would lie in between.

USES FOR RELEASING PAPERS

Pressure Sensitive Adhesives

Labels and Foams

Probably the largest single use for releasing papers is as backing for all types of pressure sensitive labels and foams. The individual requirements for specific kinds of labels can vary quite widely. The pressure sensitive adhesive is applied by means of two main techniques: (a) direct application of the adhesive to the label stock followed by drying and then lamination to the release paper, and (b) application of adhesive to the release paper followed by drying and then lamination to the label stocks.

The release requirements are more severe if the solvent solution of the adhesive is applied to the release paper because the adhesive has much greater opportunity to penetrate the paper. This method of coating the release paper is growing in popularity and is the most practical way of applying pressure-sensitive adhesive to foam at present. Pressure-sensitive labels are generally sold either in sheets or rolls. Where sheets are involved, plain sheets of various kinds of label stock pressure-sensitive, coated, and backed with release paper are sold to printers. The stock must be flat and have good printability and must feed properly on the printing press. Sometimes the label stock is die-cut into various shapes after printing. The cut is made through the label stock but not through the release paper and the "lace" is removed. Sometimes aluminum foil, metalized Mylar or other plastic films are used instead of paper as the label stock. In these cases it is extremely important that the release paper should have good dimensional stability else the stock will curl as the humidity in the air changes. Caliper uniformity is important, as is resistance to shattering during die-cutting. Extreme smoothness is particularly important when metalized Mylar is used.

Printed pressure-sensitive labels of various shapes are also offered in roll form. In these cases, the plain pressure-sensitive coated label stock is first produced. In a second operation, the stock is printed, die-cut, and the "lace" is removed all in one operation at a fairly good speed. The release paper requirements are severe and may be summarized as follows: (a) extremely good release at relatively high stripping rates; (b) good shatter resistance during die-cutting; (c) good strength because the "lace" must be removed in one piece; (d) must adhesive-coat well and have good strength at low humidity.

Another unique kind of plastic label is widely used. A colored plastic is prepared in such a fashion that when it is scored or embossed it will turn white in the scored area. There is pressure-sensitive adhesive on the back so the label or identification may be adhered to boxes, machinery, shelves, and file cabinets. The plastic is embossed while the release sheet still protects the adhesive. Here the release sheet must stand the embossing without cracking and still release readily. The paper should also be suitable for pressure-sensitive adhesive coating because it is not practical to coat the plastic. Usually the manufacturer wants to print the release paper on the back side.

Strips of foam rubber or plastic are widely used for insulating purposes by the automobile industry. Usually the foam is cast on a release paper and cured. The paper is then removed and discarded. A second release paper is coated with the pressure-sensitive adhesive and dried and combined with the foam. Good release, good performance in the adhesive coating operation, and good resistance against edge nicking or fracturing during the slitting operation are required.

Decalcomanias

In the preceding section, attention was given to the labeling technique where a base label stock was coated with pressure-sensitive adhesive or had pressure-sensitive adhesive applied to it and was later printed and die-cut. A whole new technology has developed wherein basically there is no label stock. Instead, layers of film, ink, and pressure-sensitive adhesive are laid down in successive printing steps and the final pressure-sensitive surface is covered with release paper. By such methods the need for a base stock upon which to print is eliminated. A releasing paper is first coated with a transparent tough plastic film, either over the entire surface or by means of printing in specialized areas of the sheet. In the next step, printing inks are applied to the plastic film and, in the final step, the pressure-sensitive adhesive is applied, usually by some sort of silk screen technique. The second release paper is applied over the top of the pressure-sensitive adhesive in some cases or, in others, the operation is arranged so that the pressure-sensitive surface contacts the back side of the next sheet of labels. In those instances where two release sheets are used, it is important that one of them have a good degree of transparency so that the label can be read. This type of decalcomania is used for labeling metal

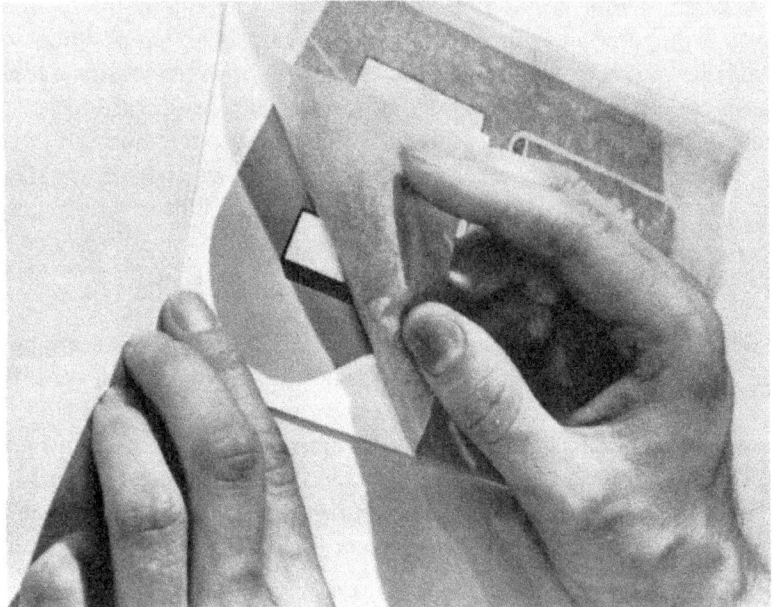

Fig. 11-8.   Pressure-sensitive decalcomanias.   Pressure-sensitive surface is up ((*Courtesy of Dow-Corning Corporation*).

drums, bottles, all kinds of cartons and containers, and furniture (Fig. 11-8).

Once again, the release paper must function well on commercial printing equipment, i.e. the paper should be flat and free from curl and should have reasonably good dimensional stability. The release requirements on the side of the paper that contacts the clear plastic film are generally moderate. On the other hand, the release requirements of the paper contacting the pressure-sensitive adhesive are fairly severe. Generally, lightweight release papers are used against the pressure-sensitive and stiffer papers, more adaptable to printing, are used on the other side. The use of pressure-sensitive decalcomanias is growing at a very rapid rate. Generally speaking, they are more economical than pressure-sensitive labels, although they are more difficult to apply. They are available in either roll or sheet form.

Plastic Shelf and Wall Covering

Highly pigmented vinyl chloride films that have been printed in various attractive designs and coated on the back side with pressure-sensitive adhesive are widely used for home decorative purposes. The plastic is flexible and is almost ideally suited for shelf and drawer coverings because it is washable and has good abrasion resistance. The pressure-sensitive adhesive keeps the product in place and makes it easy to clean. This same type of material is also used for covering walls and table tops.

The plastic film is printed and primed on the back side to improve adhesive receptivity. The pressure-sensitive adhesive is then applied to the primed side and dried and the release paper is combined with the finished product. The only purpose of the release paper is to protect the adhesive surface until use. Moderate releasing qualities are required. Some manufacturers believe that the release paper should be rigid to support the plastic and others feel that this is not necessary and simply adds to the cost. The release paper is generally printed with application instructions on the back.

A second type of product is transparent. In these cases it is necessary to print the release paper first and then to apply the releasing treatments over the top of the printing. Sometimes a protective layer of varnish is applied over the printing before the release treatment. The presence of printing inks can have an adverse effect on the release treatment if a protective layer is not laid down first.

Supported and Unsupported Pressure-Sensitive Adhesive Films

Supported Films. These consist of a web of paper, rope paper, non-woven fabric, or even Mylar, cloth, or polypropylene film that has pressure-sensitive adhesive applied to both sides. Pressure-sensitive adhesives all have flow characteristics to varying degrees. When reinforced with a web, the strength of the adhesive bond so developed is much greater.

Products of this type are interleaved with release paper that has been release coated on both sides. Generally, the release paper is used as a casting medium and must pass through an oven, usually in a horizontal position supported underneath by occasional rollers. Because it is necessary to expel the solvent, running speeds are rather low and operating temperatures are 250° to 350°F. The final product is wound on itself, hence the need for release on the back side.

At a later time the product may be unwound and the tacky side may be applied to a wide variety of surfaces. Generally, the release paper is not removed at the time of application. Often it will remain until the final fabricated product is made and sold. At the time of use, the release paper is removed and the product is adhered in place.

The paper must have good release on both sides. It should be tough so it will not break during the coating operation and it should not curl. Good dimensional stability is desirable.

Unsupported Adhesive Films. Sometimes a pressure-sensitive adhesive is cast onto release paper without any reinforcing material, or with glass fibers or other materials embedded in the adhesive mass to give some reinforcement. This type of film has far less internal strength, which causes some serious problems as far as the release paper is concerned.

The equipment for casting is generally the same as for supported films, and the strength and curl resistance properties are equally important. Again, the coated release paper is wound on itself, frequently slit into narrow rolls, and stored for prolonged periods of time before use. When the roll is unwound, it is very important that the adhesive stays on the side where it was originally cast and does not "leg" or partially or wholly transfer on unwinding.

If the release on the side where the adhesive was originally cast is slightly poorer than that on the back side, the adhesive should "stay put." Actually, good release is needed on both sides to permit complete removal of the adhesive later. If a release differential is accomplished, the release on the back side must be as near zero as possible, while the release on the cast side should be what is normally expected from silicone treated papers, or slightly less. As of this writing (1967) a complete solution to this problem has not been found. The nature of the adhesive is also important.

Strip Casting

Sometimes the pressure-sensitive adhesive is cast in parallel lines a fraction of an inch apart and later slit into narrow rolls, slightly wider than the cast strip. The "tape" is applied to all types of printed posters and displays so that they can be adhered to a wall or other surface without coating the whole back side with adhesive. The adhesives are generally moderate in tack and strength so that the poster can be removed later without leaving a residue behind.

The requirements for the release paper are similar to those listed for unsupported adhesives. Because the adhesives are not as aggressive, there is less tendency to transfer to the wrong side.

Adhesive Bandages

One of the oldest uses for pressure-sensitive adhesives is for adhesive bandages. A technique has been developed whereby the pressure-sensitive adhesive mass is compounded on rubber calenders and no solvent is used. Sometimes the textile or the plastic film is fed through the nip of the calender and the adhesive is actually applied in this fashion. If a plastic-based material is used, it usually is supported by paper upon which the plastic has originally been cast. Frequently a prime coat is applied to the plastic to improve adhesion of the pressure sensitive adhesive to its surface.

Still another technique involves the actual calendering of a pressure-sensitive adhesive mass onto the the release sheet and subsequent winding into rolls. At a later date, the roll is unwound and combined with the textile or the plastic. Sometimes a release paper of this type can be used over and over.

Some years ago it was found that pressure-sensitive adhesive surfaces characterized by extremely high gloss and smoothness had a better initial grab toward skin and other surfaces. Consequently, the release paper actually in contact with many adhesive bandages has a high gloss. Cellulose acetate or some other glossy plastic film laminated to paper is used. Sometimes the plastic film is release-coated to reduce adhesion further and sometimes it is not.

Pressure-Sensitive Gaskets

Many gaskets have a pressure-sensitive adhesive applied to one or both sides to facilitate assembly. In some cases, the pressure-sensitive adhesive is actually applied to the gasketing material prior to die-cutting and, in other cases, the adhesive is first coated on the release paper and combined. The release paper must stand slitting or die-cutting without fracturing and must have good delamination resistance. Too much release is undesirable because the paper must protect the adhesive surface during any die-cutting or handling operations.

Release Papers for Plastic Laminates

Release papers find wide use in the manufacture of plastic laminates. The types of uses may be classified as follows:

1. Release sheets for high pressure laminates, industrial and decorative.
2. Release papers for the manufacture of continuous low-pressure laminates.
3. Release sheets used in pressing low-pressure plastic films on plywood.
4. Release papers for vinyl casting or vinyl flooring.

High-Pressure Laminates

The so-called high-pressure laminates are generally composed of phenolic or epoxy resins impregnated into unbleached papers. A number of layers are assembled and the panel is pressed between two metal cauls or, in some cases, one metal caul and a release paper. However, the really substantial use for release papers in the high-pressure field lies in the manufacture of decorative laminates. Most of the decorative high-pressure laminates are composed of a core of layers of paper impregnated with phenolic resin. This core is covered with a printed white paper that has been saturated with a melamine formaldehyde resin. A final "overlay" sheet is laid over the top of the printed paper. The overlay sheet shows a high melamine resin content and becomes transparent on pressing.

A sketch showing a typical assembly for a high-pressure decorative laminate is shown in Fig. 11-9. Traditionally, high-pressure decorative laminates exhibited a high gloss. Sometimes this was dulled slightly by buffing. The high gloss finish was secured by pressing the malamine formaldehyde side against a highly polished steel caul. Two

POLISHED STEEL CAUL
MELAMINE IMPREGNATED OVERLAY
MELAMINE IMPREGNATED PRINTED PAPER
LAYERS OF PHENOLIC RESIN IMPREGNATED PAPER
RELEASE PAPER
LAYERS OF PHENOLIC RESIN IMPREGNATED PAPER
MELAMINE IMPREGNATED PRINTED PAPER
MELAMINE IMPREGNATED OVERLAY
POLISHED STEEL CAUL

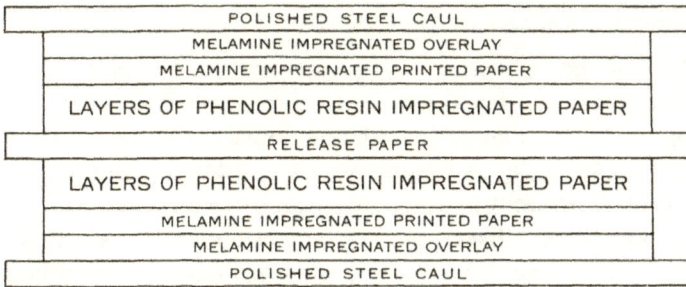

Fig. 11-9. Typical construction of a high-pressure decorative laminate.

panels were made between a set of cauls. A release paper was used to separate the back side from its neighbor. Here, good release action from the phenolic resin itself is required. A cure may run to 45 minutes at 375 to 400°F at pressures of about 1000 lb/in². For release from the phenolic side, the release paper is normally treated on one side only and a single sheet is used between two panels. After pressing, the panels can be separated. The paper is abraded off the back side of one of the two panels in a separate operation.

Increasing quantities of releasing papers are used to impart special textures or finishes to the melamine formaldehyde side of the laminate. A somewhat different release problem is presented because the melamine resins themselves are more reactive. The release paper is placed between the melamine-impregnated overlay sheet and the polished steel caul. Some interesting matte finishes can be produced in this manner.

Release papers used either on the phenolic or on the melamine side must peel readily from the plastic laminate after pressing and leave no residue or deposit behind. They must have good heat stability and should have high barrier characteristics against penetration by the melamine or phenolic resins. The surface of these release papers must be absolutely uniform, as any variation in texture will show up in the finished laminate.

Plastic laminates with a wood grain pattern made with the release paper facing the melamine overlay side look very much more like natural wood than does the same panel produced against the polished steel plate. A variety of different types of finishes can be produced by varying the nature of the release paper.

Low-Pressure Continuous Laminates

Most of the laminates in this category are fiberglass-impregnated with polyester. As far as surface texture is concerned, the situation is analogous to the high-pressure laminate field. There is an increasing interest in matte finish laminates. A substantial number of ceiling panels for fluorescent lighting are made by using a releasing paper on one or both sides of the plastic laminate at the time it is manufactured. Generally speaking, the release papers are fairly heavy because they handle better and there is less tendency toward wrinkling in the continuous lamination. The uniformity of the surface is still extremely important although the release problem is much less severe. Ordinary cellophane, uncoated, generally releases satisfactorily from polyester laminates as does vegetable parchment, plain or treated with the chromium complexes.

Some very large plastic shapes for trailers are laid up on molds and covered with a release paper and then autoclave-cured, by means of a rubber-bag molding technique, similar to that developed for plywood a number of years ago.

Manufacturers of "pre-impreg" are legion. This is either paper or fiberglass impregnated with resin. Fabricators can purchase this and lay it up on molds and then after application of suitable heat and pressure, produce a final product. Release papers are used to interleave the pre-impreg until the molding operation.

Plywood Facings

Plastic laminates have been bonded to plywood for decorative effects for years. An interesting new technique calls for application of a supported or unsupported, partially-cured, thermo-setting resin film to the surface of plywood. Quite a wide variety of polymers have been used. One type that is receiving increased attention is a polyester film, which is laid on the surface of the plywood, followed by a sheet of release paper. The panel is subjected to heat and pressure at the same time the base plywood itself is made. The final product is a finished plywood panel with good durability suitable for buffing and waxing. The end use is for paneling in construction work or, in some cases, furniture and cabinets.

Sometimes the resin film is cast directly upon the release sheet and wound in rolls and shipped in this form. It is later sheeted and inserted in the plywood press at the proper time.

Pigmented plastic films have also been used for this general application. Many give the effect of painted wood and have outstanding weathering and durability characteristics. A release sheet is frequently needed between the cauls and the plastic surface to prevent sticking, or to provide specific surface effects.

In all of these applications, the exact nature of the surface of the release paper is extremely important. By choice of the proper release paper, different surface textures can be produced. Generally speaking, the release requirements are not severe.

Vinyl Plastics

Vinyl floor tile has been produced with a matte finish through application of releasing papers. In one instance a specially treated vegetable parchment is used. The vinyl is hot-pressed. Vegetable parchment normally puts a quite rough texture on all plastic laminations, similar to that of frosted glass. However, in the case of vinyl flooring, if the release paper is removed before the vinyl has cooled and while the vinyl still has some opportunity to flow, an interesting smoothing effect takes place, A very attractive floor tile is possible by this technique.

Release papers have been widely used in large quantities as casting sheets for plastisol and organisol films. The dispersion of vinyl in plasticizer is coated onto the release paper and then fused for a few seconds at 400°F. Later, the film is removed and the paper is used over again, often four to six times before discarding. A clay-coated paper with carboxymethyl cellulose as the binder and a chromium complex release treatment has been effective in this application. The finish imparted by the clay-coated stock (weighing about 100 lb/ 3000 ft²) provides a desirable effect on the vinyl plastic. Most of the vinyl used for medical bandages is produced in this way. A similar technique is used in the manufacture of certain types of vinyl-coated fabric for automobile upholstery.

Applications in the Rubber Industry

Rubber Shipping Bags

The problem of packing and shipping crude synthetic rubber has always been an extremely difficult one. The rubbers have varying degrees of tack, depending upon the grade. All rubbers have at least some cold flow, which presents a tremendous problem in rubber packaging. The ideal unit for packaging rubber should be rigid enough to contain the product and yet should exhibit extremely low adhesion to the rubber itself and have good strength as well as economy.

The best compromise with cost developed to date consists of a multiply kraft paper bag. The inner liner is generally a special grade of silicone-coated kraft paper with good release. The Keil test is widely used as a standard for judging release. Sometimes, as an additional element of protection, the block of rubber is first wrapped in polyethylene and is then placed in the kraft paper bag with the silicone-treated parchment liner.

Problems in fabricating paper bags with silicone-treated kraft interliners arise. Stitching the bottoms of these bags since they cannot be glued satisfactorily is necessary and this in effect provides entry points for the rubber as it flows. The most common unit of weight for block synthetic rubber is 50 lb. It is essential that the block synthetic rubber should be capable of falling out of the bag when it is opened and turned upside down.

When the bags are stored at elevated temperature, problems multiply. Many times the bags are actually crated. An interliner of board to which silicone-treated kraft has been laminated provides the interior of the crate. The object of this technique is to restrain cold flow. The most popular shipping method is 50-lb bags on skids. Generally speaking, the release quality of the silicone-treated kraft is adequate for most types of crude synthetic rubber. However, the industry is still looking for a better package, particularly one that is capable of coping with the cold flow of synthetic rubbers more effectively.

A special corrugated box for packaging rubber has been developed. The inside has a layer of silicone-treated kraft paper laminated to it. The box provides much greater rigidity than does a bag but is more costly. The rubber can still flow behind the bottom flaps unless the block of rubber is first wrapped in polyethylene. Even corrugated boxes have trouble retaining their original shape after rubber has been stored in them for prolonged periods.

At one time a special clay-coated kraft paper was used in rubber bags. The clay was loosely bonded to the paper and came off quite readily, thus affording release by gross transfer.

Autoclave Curing

Much work has been done on the use of releasing papers in autoclave curing of rubber. The rubber is calendered onto the release sheet in the proper thickness and wound tightly into a roll. The roll is overwrapped with canvas strips to increase the tension and then inserted in an autoclave for curing. After the cure is complete, the roll is removed and allowed to cool; it is then unwound and the paper is stripped away. Unfortunately, most release papers entrap small amounts of air between the rubber and the paper itself. Dissipation of these air pockets is extremely difficult and would often result in a surface imperfection in the cured rubber sheeting. The problem has been to develop a release paper of sufficient porosity to provide escape of entrapped gases and yet insure adequate release and internal strength to facilitate easy removal. Special grades of smooth papers that function successfully in this type of application are available. A certain degree of wet strength is also desirable because of the condensation of steam which occurs when the roll leaves the autoclave.

Shoe Interleaver

In the fabrication of certain types of shoes and overshoes, sheet rubber, plain and reinforced with fabric, is calendered onto a releasing paper and then cut into sheets. As many as 10 or 12 layers of rubber interleaved with releasing paper are stacked and die-cut in one operation. The die-cut portions are saved for the next operation and the paper "lace" is removed from the rest of the rubber. At a later step in the operations, a stack of die-cut parts is "shuffled" with the fingers, much as a pack of cards might be broken before use. This operation is sufficient to separate the release paper from the rubber. The release paper is discarded and the rubber is ready for use in subsequent operations.

The releasing requirements in this application are quite mild. However, most papers have a tendency to shatter and break in the die-cutting operation because they do not have enough stretch to adjust to die-cutting so many layers at a time.

General Purpose Interleaver

Many applications in rubber processing and fabricating require an interleaver. Sometimes this takes the form of compounding rubber and calendering it out onto a release paper and winding into rolls. These may well be packaged and shipped to another rubber processor who carries on further operations. Release papers have long been used as liners for trays used to hold and transport formed rubber parts that are still tacky. Release papers are also used as separator sheets for slab rubber.

Core Covers

Any time that sheet rubber is calendered onto release paper and wound into a roll, the sheet has to be wrapped on something, usually a fiber core covered with specially treated release paper to prevent fiber-picking. Such cores are also widely used for the wrapping of camelback into rolls. Smaller fiber cores, covered with releasing papers, are used for winding pressure-sensitive tapes of all kinds and also supported and unsupported pressure-sensitive adhesives that have been cast first on release papers.

Caulking Compounds

Caulking compounds function as sealants in a wide variety of applications. Many sealants are extruded into place, particularly in building construction. However, in many other fields, the sealant may be extruded in advance in some particular shape so that it will fit into a specific area. Such items as refrigerators, automobiles, automatic washing equipment, and many other consumer products require sealants for certain locations in the appliance. The manufacturer extrudes the sealant in the precise shape desired. Sometimes the sealant is reinforced with wire or some other material. In other cases, it may simply take the form of a strip or a series of strips laid on a release paper and later wound into a roll.

Most sealants are characterized by high flow, at least at one stage in their life, and quite low internal strength. They frequently have substantial percentages of oil and at times can be extremely tacky.

For those caulking compounds that have quite low internal strength and are extruded in a flat strip, creped papers are widely used as interleavers. In particular, creped silicone-treated vegetable parchment has been very successful. The sealing compounds that are extruded into various shapes generally have more rigidity and greater internal strength. For these applications, flat releasing papers are much more widely used. Frequently, the width of the release paper can be quite narrow and therefore good toughness and strength and resistance to edge tearing is a desirable characteristic. Fabricators in this field usually require excellent release. Some like a glossy paper and others require good resistance against penetration by oils. Release from this class of products is difficult, chiefly because of the relatively low internal strength of most of the sealants, coupled with fairly substantial thicknesses.

Casting Sheet for Plastic Foams

Plastic foams are made in a simple form that consists of a bottom and sides, perhaps 12 to 14 in. high, which is in effect continuous. Releasing paper is used to line the inside of the form and the foam is cast directly on the paper. It rises to the proper height and then undergoes its cure. After the cure is complete, the paper can be removed without delamination of the foam itself. By proper choice of paper, wrinkling can be avoided. Polyvinyl chloride foams may also be cast in a similar manner. The foams may vary in their rigidity quite widely. The releasing requirements are not extremely severe. Most important is a fiber-free surface that is impenetrable to the ingredients in the foam. Vegetable parchment and greaseproof papers, specially treated with chromium complexes, have been widely used. Most foams are cast in widths of over 100 in. and this procedure has severely restricted the types of release paper that can be used. Generally speaking, vegetable parchments and greaseproofs offer the best results but silicone-treated krafts and certain waxed papers have also been used successfully.

Release Papers in Food Applications

There are three major areas of use for release papers in the food field:
1. Wrapper and interleaver for frozen food products.
2. Bakery tray liners.
3. Candy table and pan liners.

Release Papers for Frozen Foods

One of the largest uses for release papers in the frozen food field is as an interleaver for frozen ground meat, principally hamburger. The meat is formed into a round or rectangular patty by machine and each patty is automatically interleaved with release paper. The size is about 4×4 in. The interleaved patties are later frozen. The freezing takes place generally between 10 and 30 minutes after interleaving but there is still time for the moisture in the meat to penetrate the release paper. When the patties are used, they are separated frozen and placed frozen in the pan. (See Fig. 11-10.) The principal requirement is that the patties be easily separated, preferably by the fingers alone. Tremendous quantities of release paper are used in this way.

Fig. 11-10. Frozen hamburger patties are removed from release-treated vegetable parchment and placed in the pan while still frozen (*Courtesy of Dow-Corning Corporation*).

Two main types of interleavers are in use. One consists of two pieces of waxed paper that have been lightly sealed together along the edges. Two sheets are placed between each patty. After freezing, when the patties are separated, separation actually occurs between the two sheets of waxed paper. Silicone-treated vegetable parchment has also been used for the same application. In this case, a single sheet is used. Release is approximately the same as for the two sheets of waxed paper.

The release paper must feed satisfactorily on the equipment that forms and interleaves the hamburger patty. Such paper must have good slip and a certain amount of strength. Generally speaking, a single sheet functions better on the feeding equipment.

Release papers are used as wrappers for all types of meats that are later frozen. Good moisture barrier characteristics are also needed to prevent excessive dehydration.

Frozen cakes and pies and other pastries are becoming increasingly popular. Any of these frozen products require some sort of a release liner on top. This generally takes the form of a circle in the top of the aluminum tray in which the product is sold. The circle is almost always printed and releases readily from the frozen frosting.

Bakery Tray Liners

Release papers are extensively used as liners for pans in which all kinds of baked goods are prepared. Breads, cookies, and many kinds of sweet rolls and macaroons can be baked directly on releasing paper. This eliminates the need for greasing the pans and actually keeps the product free of the acrid taste often caused by burning grease.

The use of release papers as pan liners eliminates frequent greasing and greatly reduces washing. The pan liner may be used as many as four to seven times before discarding, still with a fresh area each time if the baked goods are properly placed.

Frequently, special folding cartons that are lined on the inside with release paper are prepared. Raw or partially baked rolls and other pastries can be placed in these containers and overwrapped with cellophane and sold. The final baking is completed in the home.

For these types of applications, greaseproof papers and vegetable parchments treated with silicones or chromium complexes give excellent results.

Candy Pan and Table Liners

Many kinds of candy can be poured hot onto tables or pans lined with release papers. The candy can be easily stripped when it is cooled. Again, greasing is eliminated. Even caramel will function in this way. The same general types of papers that are used for bakery tray liners serve equally well for candy.

Miscellaneous Uses

As can be seen from the uses that have already been enumerated, the scope of the release paper field is extremely broad. Each year, the quantity and kinds of release papers used increase. One source estimates that the market has increased at the rate of 20% per year during

1963, 1964, and 1965, with no indication of slackening. Consequently many new applications appear, all requiring the development of specific release papers to meet the particular requirements of the end use.

Some of the uses of release papers that have not been enumerated previously are: in circle form for interleaving pressure-sensitive tape, specially release-treated parchment as an interleaver for U. S. postage stamps, interleaver for tacky slab wax, liner for all kinds of asphalt containers into which asphalt may be poured hot and later stripped cold, interleaver for the packaging of special selfsealing roofing shingles, and as an interliner for ice bags to prevent sticking of ice cubes to the inside of the bag.

REFERENCES

1. de Monterey, F., Paper presented at 138th meeting, American Chemical Society, "*Silicone Coatings for Paper, Release Mechanisms.*"
2. Warrick, U.S. Patent 2,460,795.
3. Greiner, U.S. Patent 2,588,828.
4. Kather, Litster and Brown, U.S. Patent 2,803,613.
5. Dumas, Canadian Patent 550,077.
6. Dennett, Canadian Patent 556,945.
7. Dennett, Canadian Patent 557,924.
8. Iler, U.S. Patent 2,273,040.
9. Reppe, U.S. Patent 2,118,864.
10. Kung, U.S. Patent 2,381,063.
11. Kellgren, U.S. Patent 2,496,369.
12. Dahlquist, U.S. Patent 2,532,011.
13. Eckey, U.S. Patent 2,558,548.
14. Thomas, U.S. Patent 2,682,484.
15. Williams, U.S. Patent 2,829,073.
16. Webber, U.S. Patent 2,822,290.

chapter 12

Carbon and Other Impression Copying Papers

D. R. HURLEY

BACKGROUND

One of the first methods of making duplicate copies utilized a copy-writing ink. Because the copy ink had exceptional color strength, it was possible to produce additional copies by placing a thin sheet of dampened tissue against the original. These early attempts to obtain copies from an original document were indicative of the need for more efficient copying systems.

As a result of insistent demands for improved copying methods, the first carbon papers were produced by crudely coating a heavy absorbent paper with a mixture of grease and carbon black. The copies from this rudimentary formulation were extremely dirty and became badly smeared with use. In addition, the ink soaked into the base stock with age, and lost its ability to produce a copy. Both problems were largely overcome by the addition of waxes to the carbon dope. With this change in formulation the coated sheets began to resemble the carbon paper in use today.

The early manufacturers of carbon paper were highly dependent on the skill of their carbon ink formulators. Most of them had little formal training, but relied on experience gained through trial-and-error methods. The formulas they developed were jealously guarded trade secrets, and little or no information on ink formulation was published.

As the industry progressed, many new problems arose to challenge the skill of the carbon formulator. The increasing complexity of these problems forced the gradual transition from the isolated practical formulator to co-ordinated research programs that employed a wide range of scientific skills. Today's leading manufacturers of carbon paper routinely use new and sophisticated scientific techniques in their efforts to meet the many demands placed on their products.

Modern business procedures require a constantly increasing volume of written documents, In addition, more copies of each original are required to provide the necessary line of communication between departments. Despite the fact that many new methods of producing copies have been developed, carbon paper remains unmatched for its ability to provide inexpensive multiple copies under a wide range of operating conditions. Although each new copying method fills a particular need, the sales of carbon paper continue to grow. In addition to economy, carbon paper provides extreme versatility. Carbons perform in tropical heat and arctic cold; carbon images are visually, optically, and magnetically read; and carbon has even been designed for data recording in orbiting satellites.

Hence, literally hundreds of different carbon papers are sold to satisfy these requirements. All these, however, would fall under one of two broad classifications: *Single-Use Carbon* or *Multiple-Use Carbon*. Because each covers a wide variety of products, they can be subdivided into two smaller classes as follows:

> *Single-Use Carbon*
> (a)   One-time carbon
> (b)   Book carbon
> *Multiple-Use Carbon*
> (a)   Typewriter carbon
> (b)   Pencil carbon

These four classes will now be considered in detail. Special-purpose carbons will be described under each heading.

ONE-TIME CARBON

The greatest use of carbon paper is in the production of manifold business forms. These forms are gathered on special collators that attach the carbon paper to the printed sheet, and assemble manifolds with the desired number of parts. The user completes the form with a typewriter, or other imaging device, and then removes and discards the carbon paper. Since one-time carbon has fulfilled its function after one use, it is coated on the cheapest possible base paper at high speeds with a minimum deposit of low-cost ink.

Because the carbon is usually glued into the form, one must often provide an uncoated edge, or stripe, in the area where the glue is to be applied. This requirement means that the carbon must be custom-coated to match the form in which it is used. The ink formulation must also be matched to the conditions of use. The choice of formula depends upon the number of copies in the manifold, the thickness and type of the manifold paper, the method of imaging, the temperature of use, and the desired intensity of copy.

Market

As most one-time carbon is used in the manufacture of manifold business forms, statistics available on the size of the market for forms can be used to estimate the one-time carbon market.

The relative value of the carbon paper in a form varies with the type of form, but 10% is often accepted as an average value. This percentage has been applied to U. S. Department of Commerce reports[1] for 1958 through 1963 on the sales of manifold business forms, and the results are shown in Fig. 12-1. The value for 1965 was obtained from business forms sales figures obtained from other sources. Approx-

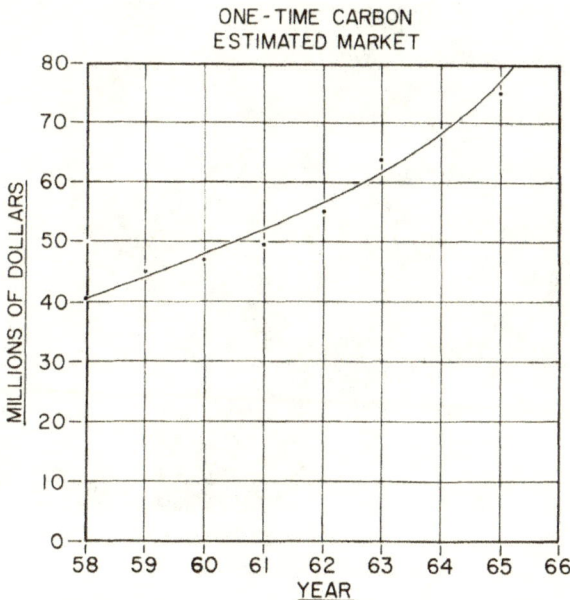

Fig. 12-1. Estimated market for one-time carbon paper.

imately 50% of this market is served by manufacturers of integrated forms that produce carbon for their own use, whereas the remaining 50% is supplied by specialists in carbon coating.

Base Papers

Most one-time carbonizing tissues are unbleached kraft, and are produced on high-speed Fourdrinier paper machines in widths up to 210 in. and at speeds as high as 3000 ft/min. This combination of wide width and high speed yields a low-cost paper. An ideal carbonizing tissue has a uniform thickness, is free from pinholes and mechanical defects, and has a surface designed to provide the proper clinch for the carbon dope. The basis weight for carbonizing paper is generally stated in pounds per ream of 500 sheets, 20 × 30 in. On this basis, one-time tissues are produced in the weight range of 5 to 12 lb. The 8- to 10-lb weights account for a large share of the total production because they provide good performance at minimum cost. The thinner papers are generally used in forms that require a large number of copies,

Fig. 12-2. Machine for making one-time carbon paper (*Courtesy of International Paper Company*).

and the heavier tissues in applications where greater mechanical strength is needed. The picture of a modern one-time paper-making machine (Fig. 12-2) illustrates the degree of technology involved in meeting today's rigid product demands.

Ink Formulation

One-time inks consist primarily of a pigment dispersed in a suitable vehicle. The vehicle for one-time ink usually consists of a small amount of hard wax, mineral oil plasticizers, and paraffin wax. Carbon black is the common pigment for black inks, whereas iron blue provides the color for blue inks. Small amounts of dyes are often used for additional color strength and improved ink flow. The following formulas illustrate the types of materials commonly used in one-time carbon inks:

Black One-Time Carbon

Ingredient	Parts
Carnauba wax	6
Montan wax	8
Paraffin wax	33
Mineral oil	25
Carbon black	15
Methyl violet	1
China clay	12
Total	100

Blue One-Time Carbon

Ingredient	Parts
Carnauba wax	10
Montan wax	4
Paraffin wax	20
Mineral oil	15
Petrolatum	20
Iron blue	21
China clay	10
Total	100

One-time inks, like most carbon inks, must be dispersed or "ground" in some manner to "wet" thoroughly the pigment particles with the ink vehicle. Many different types of heated mills have been used including stone, pot, roller, celloid, and ball. Most inks today are produced in ball mills, because they combine high output with low cost. Fig. 12-3 shows a typical ball mill of the type used for manufacturing carbon inks. Black one-time carbons have traditionally been used in forms for typewriter use, and blue for pencil or pen impressions. Black accounts for approximately 70% of the market.

Fig. 12-3. Steam-jacketed ball mill (*Courtesy of Patterson Foundry and Machine Company*).

Coating Methods

Most one-time coatings are produced on either Mayer or Flexographic coating machines. The Mayer method is the older, and employs an equalizer rod to control the ink film. The equalizer rod can be a plain steel drill rod, or a similar rod helically wound with piano wire. Wire-wound rods are generally numbered to indicate the fineness of the wire used in their windings. They are made with wire diameters as fine as 3/1000 and as heavy as 60/1000 in. Since the heavier wires apply thicker ink films, only the thinner wires are used for one-time carbon coating.

The operation of the equalizer rod in controlling the ink film is best explained by means of Fig. 12-4. Clearly the equalizer rod allows only a metered amount of ink to remain on the web. As the coating is applied in ridges, one must rely upon the leveling qualities of the ink to yield a smooth coating of uniform thickness. Ink formulations

INK ALLOWED TO REMAIN
ON SHEET

BASE PAPER

STAINLESS STEEL WIRE

DRILL ROD

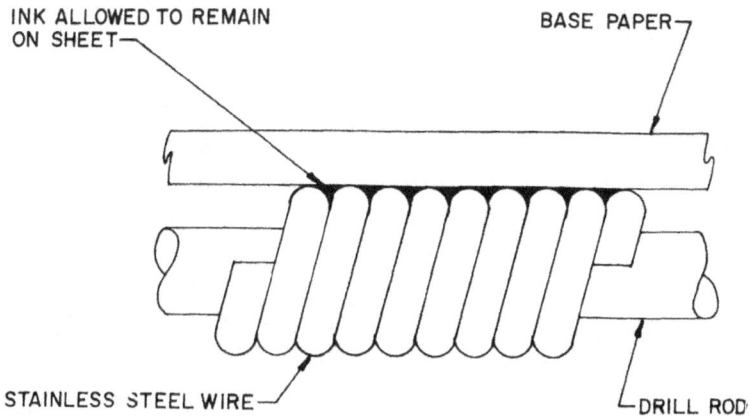

Fig. 12-4.   Equalizer rod.

for Mayer coating must, therefore, provide excellent flow.

Figure 12-5 illustrates the Mayer coating technique as it is applied
to the production of one-time carbon.   This diagram illustrates a typical roll arrangement, but many other variations are possible.   The
machine operates in the following manner.

The dope roll carries a thick layer of ink from the heated pan to the
bottom of the paper web.   The paper then passes over an equalizer rod,
which removes the excess ink.   The coated web then contacts a chilling
roll to solidify the ink and passes on to the rewind roll.   Uncoated
stripes and clean edges are provided by placing strips of thin metal
across the dope roll to prevent the transfer of dope to the paper in the
stripe area.

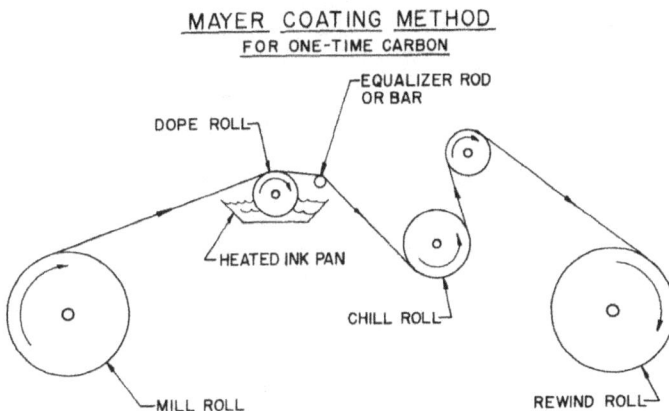

MAYER COATING METHOD
FOR ONE-TIME CARBON

EQUALIZER ROD
OR BAR

DOPE ROLL

HEATED INK PAN

CHILL ROLL

MILL ROLL

REWIND ROLL

Fig. 12-5.   Mayer coating method for one-time carbon paper.

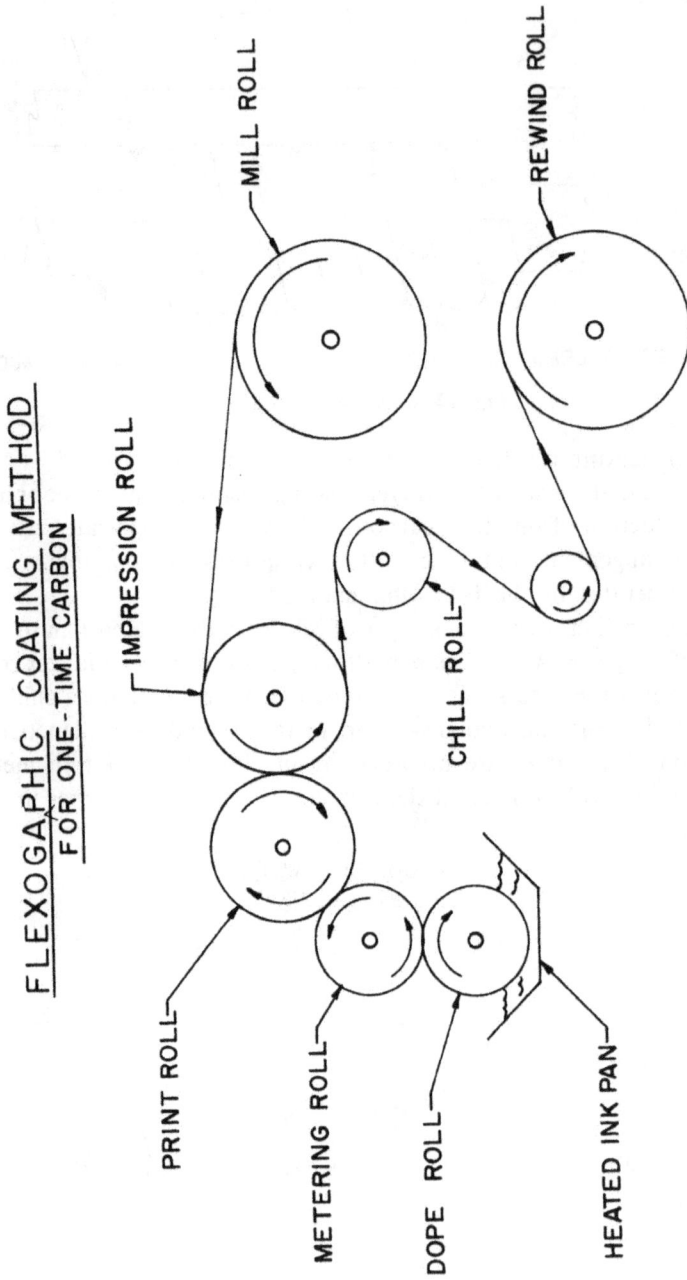

FLEXOGAPHIC COATING METHOD
FOR ONE-TIME CARBON

MILL ROLL

REWIND ROLL

IMPRESSION ROLL

CHILL ROLL

PRINT ROLL

METERING ROLL

DOPE ROLL

HEATED INK PAN

Fig. 12-6. Flexographic coating method for one-time carbon paper.

Many of the newer one-time coaters utilize the Flexographic principle. These coaters are similar to the Flexographic printing presses that have been used for many years by the printing industry. Because carbon inks are solid at room temperature, the ink pan and transfer rolls must be heated to maintain the dopes in a fluid state. A diagram of a typical Flexographic one-time coater (Fig. 12-6) provides a basis for understanding this coating method. The Flexographic coating machine operates in the following manner.

The dope roll picks up a relatively thick layer of ink from the heated ink pan and carries it to the metering roll. The squeezing action between the metering and dope rolls rejects most of the ink back into the ink pan, but allows a thin film to pass through the nip between the two rolls. This ink film is then transferred to the print roll which in turn transfers the ink to the paper as the paper passes over the impression roll. The coated web of paper then passes over a chilling roll and into the rewind roll.

Uncoated stripes can be provided by two methods. One employs a shallow groove in the print roll in the area where the stripe is desired. The other uses scrapers to remove the ink from one of the rolls in the stripe area of the ink transfer system.

Slitting

Modern one-time coaters are designed to handle large rolls of paper to minimize roll changes and thereby lower coating costs. The manufacturer of forms, however, needs much smaller rolls for his forms collators. For example, carbon can be coated in rolls 45 in. wide and 60,000 ft long, whereas the manufacturer of forms requires rolls 9 in. wide that contain 12,000 ft. The slitting of the "jumbo" rolls into smaller rolls can be accomplished on the coating machine, or in a separate slitting operation. Slitting is frequently performed as a separate operation for the following reasons: (1) it provides additional inspection; (2) defective paper can be removed; (3) strong splices can be made where web breaks have occurred; (4) Rolls of paper of the desired size can be produced without frequent interruption of the coating operation. A typical slitter-rewinder is shown in Fig. 12-7.

Processed and Spot Carbonized Carbon

Some business forms are designed to permit removal of certain sections of the carbon by tearing at a perforation. When a sheet of carbon is cross-perforated, one must keep these perforations in reg-

Fig. 12-7. Slitter-rewinder for one-time carbon paper (*Courtesy of Kidder Press Company*).

ister with the printed form. This objective is accomplished by placing along the margins a series of holes that match similar holes in the margin of the printed form. The carbon and the printed form are sprocket-fed during collation to keep them in register. Carbon paper with line-hole punching is called processed carbon.

It is also frequently desirable to have certain information appear only on specific copies of a form. For example, there may be no need to show selling price on factory copies of an order, but this information is vital for the accounting copy. This problem can be solved by utilizing spot-carbonized paper. The carbon coating is applied in a specified pattern to yield copies in only the desired areas of each copy. Here again, line-hole perforations are necessary for control of registration. Although various processing and spot-carbonizing machines have different capabilities, most will perform these operations: (1) spot carbonizing, (2) line-hole punching, (3) cross perforation, (4) linear perforation, (5) tab cutting, and (6) slitting.

Fig. 12-8 shows a diagram of a machine designed to perform all

CARBON PROCESSING MACHINE

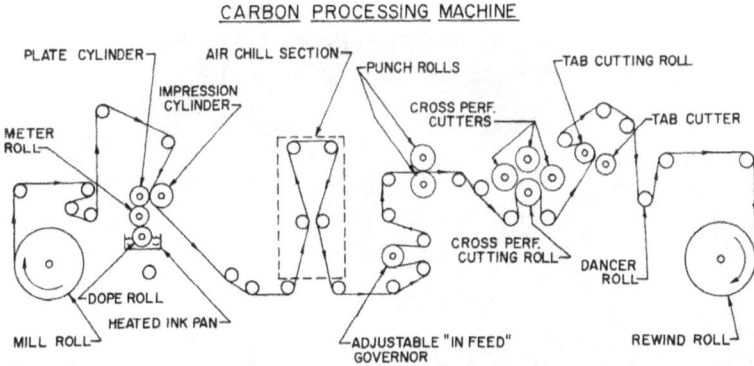

Fig. 12-8. Spot carbonizing and process machine.

these operations during a single pass through the machine. The actual appearance of the equipment to perform these functions is shown in Fig. 12-9, whereas Figure 12-10 shows the appearance of rolls of processed and spot-carbonized carbon.

Special One-Time Carbons

Many special varieties of onetime carbon have been developed for

Fig. 12-9. Typical spot carbonizer-processing machine (*Courtesy of The Hamilton Tool Company*).

Fig. 12-10. Processed and spot-carbonized paper (*Courtesy of Interchemical Corp.*).

specific applications. A description of the more important ones provides a better understanding of the diversified nature of this product.

Full or Double-Face Carbon

Full carbon is similar to regular one-time carbon except that the carbon ink is applied to both sides of the paper. This procedure results in the production of two images from a single sheet of paper. For example, a full-coated orange and black carbon is frequently used under an original designed for Diazo copying. The orange side of the paper produces an impression on the back of the original, while the other side produces a black image on the face of the following copy. Full carbon is produced by applying a coating to each side of the paper during a single pass through a coating machine. The full-coating machine is similar to the coating machine used for producing wax-backed

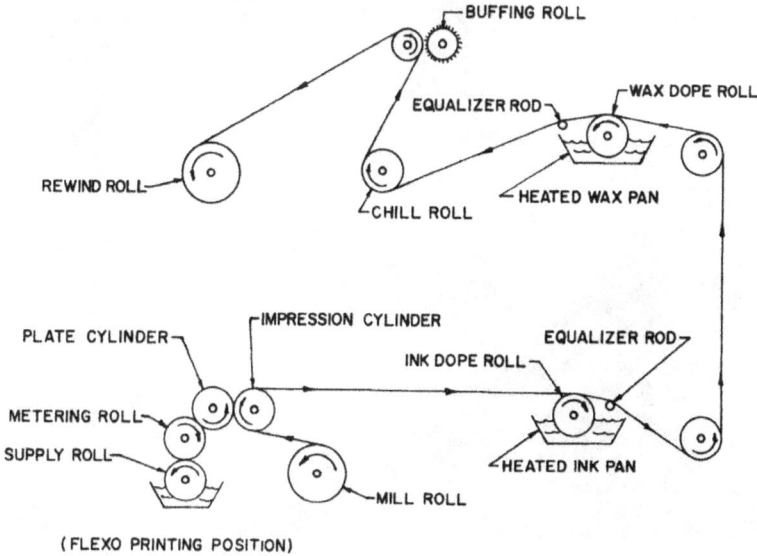

Fig. 12-11. Mayer coating method for typewriter carbon.

typewriter carbon (Figure 12-11). A special type of full-carbon paper is "saddleback carbon," which has a portion of the web coated on one side and the remainder on the other. This product is also made on a full-coating machine.

Scanning Carbon

Several electronic companies market optical scanning equipment that optically scans or "reads" properly prepared information. These scanners are extensively used in the processing of credit card sales slips. A typical credit card charge form is shown in Fig. 12-12. The bottom part of the form is generally printed on card stock which is imaged with carbon paper and eventually scanned. Because the information is used for billing purposes, it must be accurately read. Special scanning carbons have been developed that produce the image density required by the scanner, with the cleanliness necessary to minimize extraneous transfer, which can cause reading errors.

Magnetic Carbon

Optical scanning techniques suffer from the disadvantage that dirt from any source can cause reading errors. The banking industry,

Fig. 12-12. Credit card charge form (*Courtesy of Interchemical Corp.*).

after a major study, has concluded that the most practical way of getting reliable information is to utilize magnetic reading techniques. As the readers respond only to the magnetic properties of special inks, they are not confused by ordinary dirt. For additional reliability special symbols have been designed, but these have enough similarity to normal Arabic characters to permit visual reading. Fig. 12-13 shows a roll of magnetic carbon and a check that has been completely prepared for magnetic processing.

The first group of special characters near the bottom of the check indicates the Federal Reserve district, the city, and the bank. The second usually identifies the individual account number. Each of these groups of characters is printed with magnetic printing ink at the time the check is printed. The only additional information needed for magnetic processing is the amount. This information must be added after the check has been written, and is normally entered by the first bank processing the check. The entry is made with a special imprinter that utilizes a roll of magnetic carbon.

Carbon inks for magnetic imprinting utilize an iron oxide pigment dispersed in a suitable vehicle. They must be carefully formulated and coated to provide a well-defined, uniform image that is virtually

Fig. 12-13.   Magnetic carbon paper and imprinted check (*Courtesy of Interchemical Corp.*).

smudge-free under normal handling.   The quality of magnetic carbon is demonstrated by the fact that only one or two checks in each 10,000 processed are misread.

Copy Sets

One-time carbon has recently found a new and significant market in the manufacture of copy sets.   Copy sets are usually made by loosely gluing a sheet of one-time carbon to a second sheet of the type used for carbon copies.   The user places the desired number of copy sets behind the letterhead and inserts the resultant manifold in a type-writer.   After typing, the manifold is taken from the typewriter and the carbon sheets are removed and discarded.   The use of copy sets eliminates the time-consuming job of manifolding reusable carbon and the problem of storing carbon between uses.

Copy sets are produced in a number of types. The most common varieties are lightly glued, at either the top or bottom, and separation is accomplished by breaking the weak glue bond. Another common type of set is permanently glued, and the second sheets are perforated to permit separation from the carbon by tearing at the perforation. Padded, printed, and unglued sets are also available to provide the user with a wide choice of styles.

BOOK CARBON

Book carbon is a special type of single-use carbon. Instead of coating the carbon ink on a sheet of one-time tissue, it is applied to the back of the sheets that become the copies. As the carbon remains on the copy, the cleanliness of the carbon surface is extremely important. Book carbon is primarily used in the manufacture of multiple-part tags, checks, and other special forms. It is also processed into rolls for many types of accounting machines. Perhaps its most familiar use is in the manufacture of two-ply adding machine rolls. The first ply utilizes book carbon and the second ply, plain paper. Book carbon forms are somewhat more convenient to use than are those made with one-time carbon as the carbon is not discarded. In addition, they tend to be lower in cost because each sheet of paper is acting as both carbon and copy paper. Figure 12-14 shows some typical uses of book carbons.

Base Papers

Book carbon undoubtedly took its name from the book paper used as the base stock. These papers are specifically made to prevent ink penetration, and to minimize ink show-through. They are usually heavily loaded with titanium dioxide and other fillers to provide maximum opacity. For carbonizing purposes, the two most common weights are 16 lb, 20 × 30 in.—500, and 28 lb, 20 × 30 in.—500, 45 lb, 25 × 38 in. —500. The inks used are similar to conventional one-time inks, although slight modifications are made to improve cleanliness. They are usually coated on the same type of coaters used for one-time carbon.

Special Book Carbons

High-speed computers have presented a new problem to the carbon paper manufacturers. The high speed of the printer results in a rela-

Fig. 12-14. Book carbons (*Courtesy of Interchemical Corp.*).

tively light blow of extremely short duration. This blow is adequate for two or three copies, but does not produce the seven or eight copies frequently needed. A form consisting of an original and eight copies requires seventeen sheets of paper (nine sheets of copy paper plus eight carbons) if one-time carbon is used. Book carbon provides the same number of copies with only nine sheets of paper. This carbon, therefore, seems to be an ideal solution to the high-speed printer problem. They have not been widely accepted, however, because of their tendency to produce broad, poorly defined images, and their lack of cleanliness.

The carbon industry has attempted to solve this problem by developing special carbons for computer use. These are being sold under a variety of trade names and have slightly different properties. All have been designed to give sharp impressions on computer printers and to have excellent cleanliness.

These products are made in two general types—selective and non-selective. The selective papers produce an image only on specially coated receiving sheets whereas the nonselective varieties produce an image on any paper surface. Many of the products, however, fall between these extremes and have some degree of selectivity without being completely selective.

The formulations appear to be of two general types—hot-melt and solvent. The hot-melt inks are apparently based on waxes whereas the solvent products utilize resins.

TYPEWRITER CARBON

The aristocracy of the carbon paper family has been typewriter carbon. Only the finest materials and methods are used in its production. Because a single sheet can be used ten to thirty times, the cost of the carbon becomes relatively unimportant. For this reason, research has been devoted primarily to improvement in quality and not to reduction in cost.

Typewriter carbon is coated on many weights of paper with carbon inks of every hue. The backs are generally attractively printed with brand identification, and overcoated for improved appearance and curl reduction. The inks produce copy intensities that run the gamut from bare legibility to midnight black. The combination of all these factors results in hundreds of different typewriter carbons to fit the need and preference of every user.

Market

The U. S. Department of Commerce "1963 Census of Manufactures" lists total record carbon sales of $31 million per year. Other sources reveal that approximately 55% of the record carbon sales can be attributed to typewriter carbon. Applying this 55% factor to record carbon sales yields a 1963 market estimate of $18 million for typewriter carbon paper. The growth rate is approximately 3% per year.

Base Papers

The light-weight, all-rag papers made for typewriter carbon use are among the finest ever produced. They are made exclusively from rag fibers (flax, jute, and hemp) on comparatively slow and narrow paper-making machines. Few paper mills in the world have the ability to produce this grade of paper. The tissues are available in weights from 4 to 16 lb on a 20 × 30 in.—500 basis. Most are beater-dyed to produce shades of blue, purple, brown, or black.

The all-rag tissues are being slowly replaced with part-rag or all-sulfite papers in the 7- to 12-lb range. Operators of paper mills have learned to make sulfite papers of these weight ranges with performance qualities virtually equal to those of the all-rag sheets. The 4-lb papers are available only in the 100% rag category.

Ink Formulation

Typewriter carbon ink formulations bear little resemblance to those used for one-time carbon paper. Product quality is the all-important consideration and cost is relegated to a secondary position. Representative typewriter-ink and backing-wax formulations are given below:

Typewriter Carbon Ink

Ingredient	Parts
Carnauba wax	32
Ouricury wax	5
Beeswax	5
Ozokerite wax	5
Light mineral oil	26
Amber petrolatum	6
High-grade carbon black	12
Crystal violet dye	2
Victoria blue base	1
Pigmented purple toner	3
Oleic acid	3
Total	*100*

Backing Wax

Ingredient	Parts
Carnauba wax	40
Ouricury wax	40
Microcrystalline wax	20
Total	*100*

Coating Methods

Most typewriter carbon is produced on Mayer full-coating machines. These machines back-print, coat, back-coat, and buff in a single pass through the machine. The diagram of a typical machine is shown in Fig. 12-11.

The paper passes from the mill roll to the impression roll of a small Flexographic or Gravure printing press. It then travels to a Mayer coating position where a dope roll and equalizer rod apply the carbon ink. The web proceeds over a second coating position where the back-

Fig. 12-15. Typical typewriter carbon coater (*Courtesy of Haida Engineering Company*).

ing wax is applied. The paper, which is now coated on both sides, then contacts one or more chill rolls and is then rewound in roll form. A felt buffing roll is sometimes used to buff or polish the face of the sheet to improve its appearance and performance. The appearance of a commercial machine is shown in Fig. 12-15.

Special Typewriter Carbon

Typewriter carbons have remained functionally the same for many years. Recently, a new type that employs a completely new principle has been developed. This carbon produces a copy when a small quantity of liquid ink is squeezed from the cellular structure of a resin matrix. The principle of this carbon is illustrated in Fig. 12-16. The holes in the resin coating are interconnected and filled with a liquid ink. As pressure is applied locally to the sheet by a typewriter character, a small

BASE PAPER

—RESIN COATING

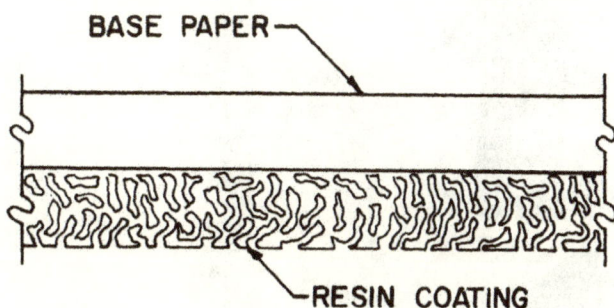

Fig. 12-16. Solvent typewriter carbon.

amount of ink is pressed from the capillaries in the image area to produce a copy. Because the ink flows freely through the capillaries, the
ink consumed is replaced from adjacent areas. This situation results in
copies of uniform intensity even with repeated use.

PENCIL CARBON (REUSABLE)

The section on pencil carbon is devoted exclusively to reusable pencil carbon, although one-time carbon is also frequently employed for
pencil use. Pencil carbons are specifically formulated to produce a
copy under the rubbing action of a pencil or pen as contrasted to the
sharp blow of a typewriter. They are extensively used in sales books,
register rolls, and peg-board accounting systems.

Market

The U. S. Department of Commerce lists total record carbon paper
sales of $31 million in 1963. If 15% of this volume is attributed to
pencil carbon, the market can be estimated at approximately $4.5
million per year.

Base Papers

The base papers used for pencil carbon vary over an extremely wide
range. Most are sulfite, with basis weights ranging from 8 to 16 lb on
a 20 × 30 in.—500 basis. The color varies widely from the white paper
used in peg-board applications, to the "red-back" frequently used in
sales books. Fig. 12-17 shows a typical pencil-carbon application.

Fig. 12-17. Pencil carbons (*Courtesy of Interchemical Corp.*).

Ink Formulation

Pencil carbons are available in two general types, oil-soluble and pigment. Oil-soluble carbons use dyes for color production, and vehicles that readily dissolve the dyes. This type of carbon is frequently used in applications that require indelibility. As the oily vehicles soak into the copy paper, the dye is carried into the paper fibers, making erasures difficult. The pigment pencil carbon formulas use pigments as the coloring agents and do not require vehicles with good dye solubility. Because the pigment tends to remain on the surface of the paper, erasures can be made with comparative ease.

Many applications for pencil carbon require each area of carbon to be used 10 to 50 times. The inks must therefore be formulated to provide a legible copy with a minimum transfer of ink. In addition, the thin layer of ink transferred must break easily from the ink remaining on the sheet to facilitate the removal of the carbon after use. Typical formulas for oil-soluble and pigment pencil carbon are given below:

Oil-Soluble Pencil Carbon

Ingredient	Parts
Carnauba wax	23
Oleic acid	15
Lard oil	16
Victoria blue base	7
Talc	39
Total	100

Pigment Pencil Carbon

Ingredient	Parts
Carnauba wax	20
Paraffin wax	7
Medium heavy mineral oil	16
Heavy petrolatum	11
Amber petrolatum	11
Toning iron blue	10
Milori blue	25
Total	100

Coating Methods

Pencil carbons are usually coated on Mayer type coating machines, which use large equalizer rods to give thick ink films. The coating conditions must be carefully chosen as the coating and chilling temperatures play a major role in determining the performance of the paper.

Special Pencil Carbons

A special pencil carbon that is similar to the special typewriter carbon described on p. 326 is available. A liquid coloring agent is squeezed out of the cellular structure of a resin matrix to produce a copy. This type of paper can normally be used at least 50 times and is extremely easy to remove from the manifold.

DUPLICATING PAPERS

Carbon paper is ideally suited for the economic production of 1 to 10 copies. Under proper conditions, one can extend this limit to 20 copies, but the legibility of the bottom copies becomes questionable. Modern office procedures frequently require 20, 50, 100, or even 10,000 copies of documents. This is the special domain of the office duplicating machine.

The three most important duplicating processes are spirit, stencil, and offset. All these methods require the production of a master with a typewriter or other imaging device. This master is then put on the proper type of duplicating machine, and the required number of copies is produced. All three methods can use a variety of papers for the finished copies, although special papers that function well are made for each method.

SPIRIT DUPLICATION

The spirit or hectograph method of duplication utilizes a sheet of special spirit carbon paper in the production of the matrix sheet. This spirit carbon is usually combined with a special master paper and is sold as a master unit. The master unit is assembled with the face of the carbon facing the receiving side of the master paper. A sheet of interleaving paper is placed between the face of the carbon and the master paper to prevent bleed of the hectograph dye into the face of the master paper. The user of a spirit master unit first removes the interleaving. An image is then placed on the face of the master paper, and the pressure of the imaging device causes carbon to be deposited on the back of the master. After imaging, the master paper is removed from the hectograph carbon and is placed on the drum of a spirit duplicating machine.

The spirit duplicator feeds the copy paper across a felt wick that is saturated with an alcohol-based duplicating fluid. The dampened paper is then pressed against the carbon image on the master sheet as the paper is fed between the master and the impression cylinder. The thin fluid on the copy paper dissolves dyes from the carbon image to produce a copy. The operation of a spirit duplicator is diagrammed in Fig. 12-18.

Spirit carbon is available in many colors, but purple and black

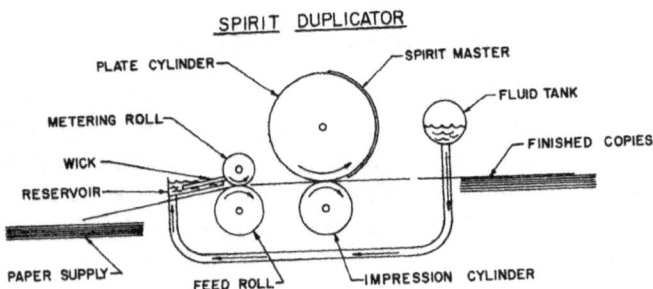

Fig. 12-18. Spirit duplicator.

are the most popular. Purple units provide up to 400 legible copies and produce maximum copy intensity. The black spirit carbons are preferred by many users although they generally produce fewer copies and tend to change color during a run.

Market

The June 1961 issue of the "National Stationers" indicates that the annual sale of spirit master paper rose from approximately $11 million in 1954 to $18 dollars in 1958. This market growth confirms the popularity of this duplicating method.

Base Papers

Most spirit carbon is coated on a sulfite book paper with a ream weight of approximately 25 lb (20 × 30 in.—500). This paper has been designed to hold the spirit carbon on the surface and to release it completely when subjected to pressure. The master paper is very similar to the carbonizing base stock, but has a ream weight of 38 lb on the same weight basis. This paper is designed to accept the carbon image from the spirit carbon, and to provide good mechanical strength after repeated exposure to alcohol-dampened copy sheets.

Spirit carbon inks must be formulated to contain large amounts of alcohol-soluble dye. Purple spirit carbons usually contain approximately 50% of crystal violet, which is dispersed in the ink vehicle. The formulas below illustrate the type of ingredients used.

Violet Hectograph Carbon Ink

Ingredient	Parts
Carnauba wax	7
Ouricury wax	5
Blown castor oil	7
Petrolatum	13
Mineral oil	14
Crystal violet	54
Total	*100*

Black Hectograph Carbon Ink

Ingredient	Parts
Crystal violet dye	8
Brilliant green dye	5
Erio green dye	5
Chrysiodine dye	21
Bismarck brown	21
Carnauba wax	13
Amber petrolatum	7
Castor oil	2
Light mineral oil	18
Total	*100*

This black formula illustrates the difficulty of making a good black spirit carbon. The black "color" is simulated by a mixture of three dyes. If these dyes do not have perfectly matched solubility rates, some will be consumed faster than the others and so cause a color shift in the copies during the run.

Coating Methods

Spirit carbon is usually coated on Mayer coating machines. The type of image obtained is highly dependent upon the crystal structure developed in the carbon dope as it solidifies. For this reason, spirit carbons are generally coated at extremely slow speeds, and subjected to controlled chilling conditions. In addition, they are usually aged for several weeks to provide time for the proper crystal structure to develop.

STENCIL DUPLICATION

The master for stencil duplication consists of a special sheet of tough fibrous paper that has been coated with a composition that is impervious to the stencil ink. The image to be reproduced is stencilized onto the stencil sheet by a typewriter, hand stylus, metal addressing plates, or other means. The stencil is then placed over the cylinder of the stencil duplicating machine. This cylinder is hollow and partially filled with a duplicating ink. The ink is fed through holes to a soft ink pad that covers the cylinder wall. The ink from this pad is forced through the holes in the stencil as the copy paper is fed between the stencil and the impression roll. The copy paper absorbs the ink to produce the copy. A single stencil can produce as many as 10,000 copies. The

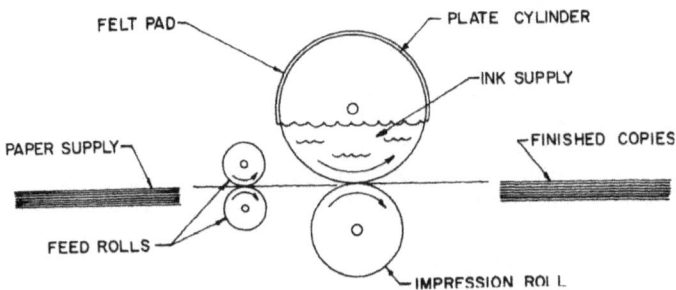

Fig. 12-19. Stencil duplicator.

color of the copies is determined by the color of the ink in the machine. Fig. 12-19 illustrates the method of operation.

A stencil assembly consists of a stencil sheet, a backing sheet and a cushion sheet. The backing sheet provides an impact surface of the proper hardness for ideal stenciling. The cushion sheet is placed between the stencil and the backing sheet when the stencil is prepared on a typewriter or other mechanical writing instruments. These cushion sheets are available in various weights, the heavier sheets yielding a broader image. In addition, a thin sheet of pliofilm is frequently placed on the top of the stencil assembly as it is typed. This arrangement provides a broader copy, eliminates letter cut-outs, and keeps the type clean.

Stencils are available in a range of qualities, to provide the desired number of copies. Most stencils are printed with a series of guide lines to aid the user in obtaining an attractive layout. Stencils are also available in a variety of colors to suit the personal preference of the user.

OFFSET DUPLICATION

The usual offset duplicator is actually a small lithographic printing press. Most machines can utilize either a metal or a paper master plate. The metal plates are generally prepared by commercial plate makers, but the paper plates are easily prepared with a typewriter and a special typewriter ribbon. The paper plates are designed to give as many as 10,000 copies, whereas metal plates can produce much longer runs. The copies produced by this method are generally of higher quality than those from the other method described.

The paper plates are produced by placing a chemical coating over the surface of a strong base paper. This paper is specially designed to withstand repeated exposure to water without losing its mechanical strength. The surface coating is hydrophilic and, therefore, easily wet by water. The image placed on the master is hydrophobic, or water repellent. Consequently the prepared plate has a water-repellent image on a water-receptive background.

This plate is placed on the printing cylinder of the duplicating machine. The machine is started and the water distributing system engaged to wet the surface of the plate. The water uniformly coats the entire surface with the exception of the hydrophobic image areas. The ink distributing system is then engaged to feed a supply of ink to the printing surface. This ink is not picked up in the dampened back-

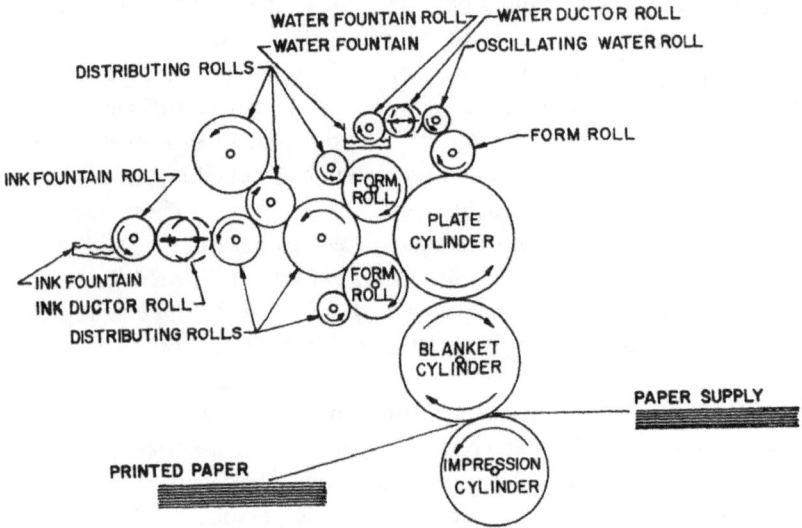

Fig. 12-20. Offset duplicator.

ground areas but is readily transferred to the image areas. The ink is then transferred from the image areas to the rubber surface of the blanket cylinder. The blanket in turn transfers the image to the copy paper as it is fed between the blanket and the impression cylinder. The color of the image depends solely upon the color of the ink that is fed from the ink fountain. Fig. 12-20 illustrates the operation of a typical offset duplicating machine.

CARBONLESS PAPERS

Many so-called "carbonless" papers have been introduced within the past several years and most have found applications where their generally higher cost could be justified because of their cleanliness. No attempt is made to describe every system, but rather to indicate the general operating principles.

Carbonless papers can be considered under the ' ᵔ broad classifications of physical and chemical process papers. The physical process sheets produce their copy as a result of removing, displacing, or compacting a layer of material to reveal the color of an underlying surface. The chemical process sheets form an image as a result of a chemical reaction. The total annual market for carbonless papers is estimated at approximately $40 million.

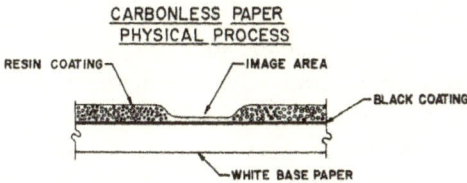

Fig. 12-21. Carbonless paper, physical process.

PHYSICAL PROCESS

Fig. 12-21 shows the operating principle of one type of physical process carbonless paper. The white base paper is coated with a black ink and overcoated with a special resin formula. This resin layer is produced in such a manner that it contains a multitude of small air bubbles. Since the air in the bubbles has a different index of refraction than the resin, a white opaque film is produced. When pressure, as from a typewriter character, is applied to the resin layer, it collapses and coalesces forming a thinner transparent layer in the image area. When the sheet is viewed, it appears white except in the area where the black undercoating is visible through the compacted resin.

The principle is perhaps better understood from an analogy. If a piece of transparent ice is powdered, it appears white because of the difference in index of refraction between the ice particles and the trapped air. If this crushed ice is subjected to pressure sufficient to eliminate the air, it again becomes transparent.

CHEMICAL PROCESS

The chemical process sheets are of two general types, which are frequently designated as single-and double-coated papers. Each type will be briefly discussed.

The first chemical process sheets placed on the market employed the two-coating system. It is necessary for two separate and dissimilar surfaces to be in contact and pressure exerted before an image can be created. The underside of the transferring sheet has a chemical coating. This coating contains capsules of colorless dyes. The front of the receiving sheet has a clay-like coating. When these coatings are in contact, and the pressure of handwriting, typewriter, or machine printing is exerted, the image appears.

The following three types of paper are necessary to meet the requirements of forms systems:

CB (coated back)—the first sheet in a set

CFB (coated front and back)—the intermediate sheets in a set

CF (coated front)—the last sheet in a set.

These three types of paper meet the requirements of virtually all forms systems. To create a two-part set, a CB sheet and a CF sheet are used. For a set of over two parts, the first and last sheets are CB and CF, and the intermediate sheets are CFB. In addition, the entire form set can consist of CFB, as this sheet has the ability to receive and transfer an image. Fig. 12-22 illustrates the use of these three types of paper.

The single-coating type of paper has all the reactants incorporated in a single sheet. One needs to stock only one type of paper if this system is used, and there is no danger of assembling forms with the improper orientation of coated layers. This system, however, suffers from the disadvantage that the sheet always retains its pressure sensitivity, even after it has been removed from the form. This property

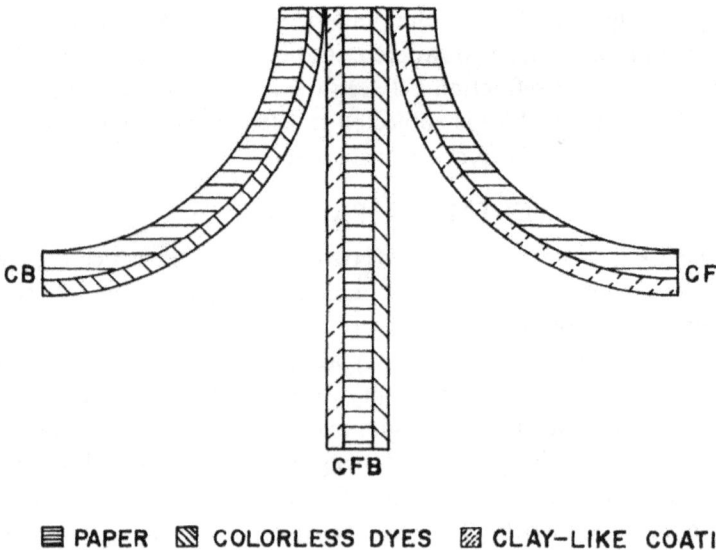

▤ PAPER ▨ COLORLESS DYES ▩ CLAY–LIKE COATING

Fig. 12-22. Carbonless paper, chemical process (*Courtesy of The National Cash Register Company*).

has limited the usefulness of the single-coated or self-contained type of sheet.

REFERENCES

1. U.S. Bureau of the Census, Census of Manufactures, 1963—*Industry Statistics: Commercial Printing and Manifold Business Forms, MC63(2)-27B*, U.S. Government Printing Office, Washington, D.C. (1966).

chapter 13

Functional and Protective Packaging Papers

HUGH E. LOCKHART

A package performs one or more of three functions: protection, utility, and motivation. As analysis of packaging becomes more sophisticated it becomes more and more evident that probably every package performs all three functions to a greater or lesser degree. The *protective* function refers equally to protection of package contents from the environment and protection of elements in the environment from the package contents. For example, reagent grade concentrated sulfuric acid should be protected from contamination by water, while persons and objects need protection from the acid. *Utility* refers to situations in which the package makes the product easier to use, move, or store. Overwrapping and palletization and containment of goods are examples. *Motivation* refers to inducing the customer to buy at retail. At other levels it can refer to package design or construction, which induces one to use the product more safely or more economically.

This chapter deals mostly with the protection function, for it deals with those properties and features of packaging papers that provide protection to and from the contents of the package. To the extent that heat seals are looked upon as necessary inhomogeneities in the barrier, they too are characteristically a part of the necessary considerations in the protective function. In performance of the protective function, papers are used as wrappers over objects or materials to prevent water, grease, vapors, or chemicals from entering or leaving the unit package. Because paper is not itself a barrier to these things, the continued success of paper as a protective material evidently hinges upon the use of laminations and coatings and upon continued improvement of the coating and laminated materials as well as in methods of applying them.

FUNCTIONAL PROPERTIES

Functional and protective papers can be classified according to their properties as follows:

1. Water-resistant	6.	Water-vaporproof
2. Waterproof	7.	Chemical-resistant
3. Grease-resistant	8.	Gas-resistant
4. Greaseproof	9.	Gasproof
5. Water-vapor resistant	10.	Heat-sealing

In the industry today, these terms usually cover relative properties and little differentiation between "water-resistant" and "waterproof" papers, for example is made. Many different tests have been devised, recommended, and used, but at present no set of laboratory tests can duplicate actual usage requirements. The subject will be discussed from the standpoint of general requirements for protection against water, water vapor, grease, chemicals, and gases. Heat-sealability or pressure-sensitivity are specific properties that can be built into coatings.

Water-Resistant and Waterproof Papers

Water-resistant and waterproof sheets have a great many applications. The degree of resistance to the passage of water depends on the usage to which the sheet is subjected. In general, there are three types of requirements:

a) Resistance to staining and absorption in the sheet by a film of water under negligible pressure.

b) Resistance against the penetration of water or aqueous solutions under a substantial head or pressure for a given time.

c) Absolute resistance to penetration of water or aqueous solutions under a substantial head or pressure.

To develop a sheet that will do the job required, yet not incorporate an expensive and unnecessary degree of resistance, the application should be analyzed so as to determine the proper characteristics needed in the sheet. The type and thickness of coating or laminant must be adequate, yet even the best coated films have a certain number of pinholes or voids, and the degree of water resistance is a function of their number and size, of the thickness of the coating, and of the hydrophobic characteristics of the surface. Where film thickness of 0.001 in. and one coating is adequate, it is certainly not economical to apply a 0.002-in. film requiring two coatings, The same is true of the required load of laminant, or the use of a specific type and thickness of a laminated film or foil. The composition of the coating is an important factor because an easily-wetted surface permits the water to seek out any pinholes or apertures in the film, whereas a hydrophobic surface, such as vinyl film, makes wetting and penetration extremely difficult if not impossible for water unless under pressure. In the development of papers that are resistant to the penetration of water or aqueous solutions, however, the problems are generally not overly complex as water is a simple compound and easily repelled, and the only major factor that must be guarded against is the presence of some surface-active agent that can cause wetting and penetration of the sheet where plain water is easily resisted.

The question of flexibility of the coating must also be considered, because a sheet built with a 100% safety factor in the flat form, but which cracks on folding, is not much better than a sheet with a 50% deficiency in the over-all makeup if an envelope or package must be fabricated from it. The development and testing of these papers requires fine judgment and the proper test conditions are determined only by experience and experiment.

Grease-Resistant and Greaseproof Papers

The development of sheets resistant to the passage of "grease" presents a problem almost diametrically opposite to that posed by water-resistant papers. Grease is hydrophobic in character and is a common term for anything from heavy motor oil to light machine oil containing special penetrants, and extending to chocolate, peanut butter, and lard. Instead of a single chemical compound such as water, the technician is faced with an extensive range of chemical compounds and a wide range of physical states including light fluids, viscous oils, and solids. This complexity of composition has forced the technician to evaluate his specific sheet on the material to be packaged in it. Most of the industry is now carrying out its evaluation programs in this manner. The various technical associations still base their evaluation tests on torpentine, however, because early specifications of grease-proofness were set up in terms of this compound. This test, although it offers a good control mechanism, does not necessarily predict the resistance the paper will offer to anything besides turpentine. Although this test is still widely used both in evaluation and control work, it is being superseded by more accurate and specific methods.

A few general aspects that may be discussed in connection with grease-resistant sheets are concerned with the nature of the transmission of grease or oil through such sheets. Cellulose itself is a good grease-resistant barrier because of its hydrophilic character, so grease or oil penetrates a sheet of paper through the interstices between the individual fibers or along the fibers themselves by capillary action and not through the cellulose. If the sheet can be so altered that there are no intersticial spaces as in the case of glassine and parchment, it will be resistant to the passage of grease or oil. The use of cellophane, or regenerated cellulose film, also illustrates this point. If a film, foil, a coating, or a laminant is used to produce the barrier, it should be insoluble in the specific grease or oil being tested. If a coating or laminant is used, it should be applied to a smooth base sheet, because any voids or apertures in the film may permit transmission and any fibers that protrude through the surface may act as wicks and pass the oil or grease through the film to the opposite side.

Three types of requirements can be distinguished:

a) Resistance to training and absorption in the sheet by a film of grease or oil under negligible pressure.
b) Resistance to penetration of grease or oil through the sheet under a substantial pressure for a given period of time.
c) Absolute resistance to any penetration of grease or oil over long periods of time and under substantial pressure.

When selecting a sheet for a certain application one should study the test methods that will be applied to the sheet and also its use requirements.

A very important consideration is how long the barrier must resist the specific material being packaged in it. A perishable food product with a greasy surface, such as doughnuts, may require a sheet possessing only a few days of grease resistance to lard or fats. A lubricating-oil container, on the other hand, might require absolute resistance to the penetration of both hydrocarbons and aromatic oils for a period of 1 or 2 years. A further consideration is the economics of using various alternate barriers for any application. The failure of a greaseproof sheet used on a machine part might not be too serious as a messy package could be the only consequence other than the loss of oil from the part because of its absorption by the package. The failure of a doughnut package would result in an unsalable product, however, and the

failure of the oil container could well result in a loss of its contents.
The flexibility of the sheet must also be considered.

There are at least three methods of grease penetration:

a) Penetration through pinholes or apertures in the sheet. The viscosity of the material determines the rate of penetration for a given sheet.

b) Wick action along and through protruding fibers.

c) Solution of the grease in, and consequent diffusion or transmission through, the sheet. The chemical constitution of the grease or oil determines these features for any given sheet.

From these considerations one can assume that the maximum grease resistance is obtained when:

a) The barrier film is continuous and free from voids, apertures, or pinholes.

b) The surface of the paper under the coating or laminant is smooth so that no fibers or a minimum of fibers protrude to act as wicks. This condition is, of course, eliminated if the barrier is a film or a foil.

c) The components of the film are not soluble in and do not dissolve the grease or oil.

Water-Vapor- and Gas-Resistant and Water-Vapor- and Gasproof Sheets

There is no absolutely vaporproof barrier since even metal foils that are known to be free from pinholes and apertures transmit minute amounts of water vapor because of the nonhomogeneity of their structure. Water vapor or other gases may also be transmitted through a homogeneous plastic film, if minute amounts can be absorbed by the surface structure, passed through by an absorption-diffusion mechanism, and then desorbed on the far side of the film. The amounts are usually small and depend upon the size, polarity, and structure of both gas molecule and the barrier, but this occurrence does illustrate the problems that must be faced in this application.

The transmission of water vapor and gases is a function of several factors including the over-all temperature of gas and of the barrier; the type, thickness, composition, and crystallinity of the barrier; and the area exposed to transmission.

Vapor-resistant papers are used for wrapping products to maintain their original moisture or gas content as long as possible. Some frozen meats may have a tendency to dry out in storage, causing "freezer burn" through moisture diffusion out of the package because of the low moisture content of the low-temperature air outside the package. Dry-packaged cereals, flour, and hygroscopic salts may tend to cake if they absorb moisture and moisture must be kept out of the package. Some foods, such as spices and essential oils lose their flavor if the aromatic odors evaporate and must be packaged so as to retain their odor and flavor. Certain foods require that the package "breathe" moisture, carbon dioxide and oxygen. Some foods are packed in nitrogen or are vacuum-packed and the specific gas must be retained or the air kept out. In many of these applications, the gas pressure is equal on both sides of the barrier and in others a pressure difference is maintained. In some cases the temperature is normal and in others it is low or excessively high.

The contents of the package may have a tendency to dry out or lose a volatile ingredient, or may tend to pick up moisture, depending upon their nature and on the conditions on either side of the barrier. If the equilibrium vapor pressure of the contents is higher than the vapor pressure outside the package the passage of vapor will be outward. If the opposite conditions exist, the passage will be inward. On the inside of the package, the vapor pressure depends on the temperature and the composition of the contents. On the outside it is a function of the temperature and, in the case of moisture, on the relative humidity.

In setting up specifications for a vapor-resistant barrier, careful consideration must be given to the nature of the contents and the equilibrium vapor pressure, the size of the package and exposed surface area, the critical amount of gas or moisture that can be gained or lost, the length of time that the material will be in the package before being used, and the temperatures at which the package will be stored. The test conditions for the evaluation of such a package are a matter of fine judgment and should be decided upon only after careful consideration of the application. It is not of much value to test a frozen food package at room temperature, for example, or to evaluate a package for use in the tropics under test conditions of 73°F and 50% relative humidity.

One specific problem involved in the utilization of vapor barriers is concerned with flexibility and folding properties. When a sheet is folded, either 90° as in fabricating a container or box cover, or completely back upon itself, as when making an envelope, intense stress and stretching occur at the fold. In some films such stretching causes

TABLE 13-1

Water-vapor Permeability of Common Films and Papers

	Thickness, in.	Permeability*
Lacquered glassine	0.0008–0.002	3–15
Waxed glassine	0.0008–0.003	3–15
Pliofilm	0.0008–0.002	8–15
Moistureproof cellophane	0.0009–0.0017	3–9
Saran film	0.001–0.002	3 or less
Polyethylene	0.004	3.8

* Reported as grams of water transmitted per $100/in.^2/24$ hr at $100°F$ and 95% relative humidity.

crystallization and in some cases a partial rupture or at least a thinning of the film results. This could obviously cause an increase in the vapor transfer rate at the point of the crease if any diminution of the film thickness took place, although if crystallization and no rupture occurred in all probability the sheet would become more resistant to passage of vapor at that point. Because most packages contain a number of such potential danger spots, the safest technique is generally to evaluate the package that will actually be used in making recommendations for any specific application.

Chemical-Resistant Sheets

In the packaging field there are many applications where the barrier must be resistant to specific materials—such as foods, chemicals, drugs, and industrial products. The requirements vary so much, however, depending upon the specific material involved, that no attempt is made here to make any analysis of individual applications. The coating must be tested in contact with the material that is to be packaged and under conditions as close as possible to those met in actual use. Some general conditions, however, are as follows:

a) Alkalinity or high pH. In packaging salts, some drugs, soaps, and other alkaline materials, the portion of the barrier that comes into contact with the contents must be resistant to alkali and not be affected by it or change its properties because of any reaction with it.

b) Acidity or low pH. For packaging certain salts, drugs, or foods that are acidic in nature, the barrier contacting the material must be resistant to weak and preferably also to strong acids and not be affected by the specific compound being packaged.

c) Solvents. The packaging of any compound containing solvents such as nail polish, drug tinctures, liquors, and cleaners invariably presents a problem as most plastic films and synthetic resin coatings and laminants are softened, if not completely dissolved by many solvents. Foil, of course, does not present this problem. Any compound containing solvents must be evaluated against the coating to be used because traces of a solvent other than that tested often affect the coating.

Some general requirements for sheets to be used with reactive compounds can be set up:

a) The sheet should not react chemically with the packaged product.

b) It should not contain any mobile component that could migrate into the packaged product. It should also be formulated so that it cannot absorb any component of the packaged product.

c) It should not impart any odor, particularly objectionable ones, to the product.

d) If the product is used for food, drugs, or any compounds to be taken internally, the barrier must be absolutely free from toxic components that could contaminate the product.

e) If the packaged product is of the type that contains a critical amount of moisture, the coating must be sufficiently water-vapor resistant to maintain it in the proper condition. An example of this would be citric acid, which in the dry crystalline form would be easy to package. If the contents were in a concentrated solution of citric acid, it might attack and destroy the barrier.

Heat Sealing Papers

These papers have the property of being self-sealing under the combined action of heat and pressure. The purpose of such a seal is to make in a package a fused closure that is as resistant to grease, water, and water vapor as is the surface area of the sheet itself. A heat-sealing coating is necessarily thermoplastic, but it must also be sufficiently nontacky and nonblocking so that the face will not block to the back in the roll or in skids. Another problem with this type of sheet is the tendency for the two coated faces, at spots other than at the heat seal, to block because the sheet made into an envelope or sleeve container must be capable of being packed flat for shipment; if the envelopes cannot be opened to insert the product to be packaged, they will have no value.

The mechanism of heat sealability depends upon at least five factors:

a) The temperature of the sealing platen.
b) The pressure applied on the seal.
c) The time of contact at the designated temperature and pressure.
d) The heat transfer characteristics of the barrier.
e) The softening characteristics of the heat sealing film.

When the coated sides of the two sheets, or of one sheet and an uncoated surface, as in a lap seal, are pressed together and a heated surface is applied to one or both surfaces, heat transfer to the coating takes place. The coating film either melts or becomes viscous and tacky, depending upon the type of coating. A paraffin-wax coating, for example, melts at 125 to 145°F, depending upon the grade of wax and the composition of the coating, whereas a coating formulated from a cellulose derivative or a vinyl resin becomes viscous and tacky at 185 to 275°F. A coating that has a sharp melting point and which melts to a low-viscosity solution is usually not sticky while in this condition, with the result that it is not able to bond the two sheets together until it is chilled and becomes solid once more. When this type of heat-seal coating is used, the two sheets must be held together after contact with the heated platen and preferably in contact with a cooling mechanism until the temperature of the seal drops below the solidification temperature so that a bond can be set up. With a resinous coating, on the other hand, where the coating becomes viscous and tacky when heated, the two layers fuse and the cohesion between the layers is usually sufficiently strong to bond them and hold them together even at the temperature at which the sealing takes place.

The temperature to which the two layers of coating must be raised to bring about activation of the heat seal and a resulting flow and tackiness so that fusion can occur depends upon the formulation of the coating. For different coating compositions the temperatures may range from 150 to 350°F. For any particular coating, there may be a range of 25 to 50°F over which satisfactory seals can be made. It must be kept in mind that the lower the minimum temperature of fusion, the more subject the coating is to blocking, unless a compound is found with a sharp melting point that changes the film from a solid to a viscous and tacky melt. The higher the minimum temperature of fusion, the higher must be the heating surface temperature and this would probably lead to a reduction in the speed of heat sealing.

The temperature of the heated surface of the platen must, of course, be higher than the fusion temperature of the coating so that a temperature gradient can exist for heat flow to the coating through the sheet.

The capacity of the heating unit should be sufficiently high, particularly on a continuous heat-sealing machine, so that this temperature can be maintained at the sealing speeds at which the machine is operated. A number of different machines are available; some heat the sheet while the platens are held under sealing pressure, and others supply the sheet with increasing increments of heat as the package moves through the unit until the pressure area is reached and the seal is made. Where fluid melt-type coatings are used, heat and pressure are applied to produce the seal. The heat is then removed and the seal held until its temperature is brought below the solidification point.

Just as there is a lower temperature limit below which fusion will not take place, there is also an upper limit above which either of two undesirable conditions can occur. First, if the temperature is above a certain critical limit, the heat may scorch the paper. Scorching affects its strength and may also discolor the sheet. Second, there is a limit, above which the coating will become so fluid that it will squeeze out and not possess sufficient tack to hold the two surfaces together unless they are rapidly chilled. When the coating becomes too fluid, it may strike through the paper to cause a discoloration on the outside of the sheet and it may also foul the sealing platens. Where impermeable sheets such as film and foil are being sealed this strike through will not occur, because such barriers will not pass the melt, even at its lowest viscosity.

The pressure applied is important. The higher the pressure the more intimate contact there is between the heating surface and the paper, and this situation provides for more rapid and uniform heat transfer. Increased pressure also helps the two layers of coating to flow into each other and form a fusion weld. The pressure required is a function of surface geometry, the melting point of the seal coating, and the viscosity characteristics of the seal coating. Once a good seal is achieved further increase in pressure has no effect. It neither improves nor degrades the quality of the seal, provided there are no special conditions that lead to extrusion of the seal coating from the interfacial area of the seal. Contrary to some opinion, it is impossible to obtain perfect contact between heating surface and paper. An insulating air layer is always present.

The required time of contact is a function of the applied temperature and pressure and decreases as the temperature and pressure are increased.

The temperature, pressure, and time relationship for any specific

formulation must be determined by experiment, aided by the knowledge of the type of heat-sealing equipment available for the specific
application.

Assuming that a satisfactory weld or fusion of the two layers of
coating has been effected, the strength of the seal depends upon two
conditions:

a)   The cohesion of the two layers of coating to themselves.

b)   The adhesion of the coating to the paper, film, or foil base.

If the cohesion of the coating to itself is good and the bond to the
sheet is poor, the seal may separate from one or both of the plies. If,
on the other hand, the bond to the sheet is good, and the cohesion of
the coating is poor, the seal may fail in the coating itself. Obviously
both must be satisfactory to obtain a good seal, and the entire seal
can only be as good as its weakest component.

COMBINATION PAPERS

There are many highly complex situations in the present scheme of
packaging for which no single paper will perform adequately. Paper
will not heat-seal. Paper is no barrier at all to water, water vapor, gas,
or grease. Parchments and glassines provide good protection against
grease and gases, but they are not resistant to water vapor. In addition, the manufacture of glassines necessarily severely lowers the tear
resistance of the sheet. Because of these and similar considerations,
common practice is to combine many kinds of paper with other materials to extend the usefulness of both the paper and the other material.

To any combination, paper provides a low cost, strong substrate.
On this substrate can be put more expensive materials with special
properties, in small enough amounts to be economical. The paper also
provides to these materials the opportunity for many kinds of decoration, forming, and shaping that would otherwise not be available.
The companion materials bring to paper primarily two classes of improvement: barrier properties and heat sealing capability. The combination is made in one of two ways: coating or lamination. Often a single
combination involves both coating and laminating. These combinations are not restricted to two materials, but may have four different
layers, involving two, three, or four chemically different compounds.

Coating Compounds. Processes

Coating processes involve applying a special-property material in
some fluid form so that it flows over the surface of the paper substrate.

The fluid form may involve simply melting as is the case with waxes and wax blends. Other materials are dissolved in suitable solvent and solidification depends upon various means of solvent removal. Vinylidene chloride is applied from an emulsion, and the polyolefins are applied by extrusion, which is a sophisticated way of melting polyethylene without oxidizing it excessively.

Coating materials provide barrier properties for a combination in nearly all cases. All the coating materials provide a heat-sealing capability to the paper substrate. In most cases the same long chain, olefinic or vinyl structure is responsible for both the sealing and barrier behavior of the material. Developments in coatings can likely be explained or predicted on this basis. One of the newest and most active entrants into the coatings field is the modified wax or hot melt. Wherever encountered, the newest of these are some combination of paraffin and/or microcrystalline wax with ethylene vinyl acetate copolymer. The combination provides a wax with higher melting point, greater flex resistance, and improved barrier properties. As it is a two- or three-component system, a great many useful variations in properties can be obtained by formulation change. Throughout the range of formulations, the copolymer adds a high degree of toughness to the wax. Thus, the wax application becomes much more resistant to cracking during normal fabrication and handling of packages made from the coated material.

Ethylene vinyl acetate copolymer (EVA) is manufactured by a number of concerns. Three of them are E. I. du Pont de Nemours with trade name *Elvax*, Union Carbide with trade name *Comer*, and Dow Chemical Company with trade name *Zetafax*. The types, formulations, and structures differ somewhat as would be expected, so any general statement about the material may not be quite true of any particular one.

Another of the newer barrier coatings for paper is vinylidene chloride applied from emulsion. It is reported as *Saran* type, but this is a trade name. The same material is available to suppliers as *Daran* from W. R. Grace Company. Although this polymer is an excellent barrier in sheet and film form, it has proved to be somewhat less effective than waxes as a paper coating material. The problem has been one of producing coatings so that the paper fiber does not protrude through the coating and thus wick water through it. Several coating combinations are included in Table 14-2 to illustrate their sealing and barrier characteristics.

TABLE 13-2
Sealing and Barrier Characteristics of Certain Coated Papers*

Coating material	Base material	Wt. lb/ 3,000 ft²		Heat-seal range, °F, 0.5 sec. dwell	WVTR† g/m²/24 hr, uncreased
		Coating	Base		
Paraffin	Sulfite	10	25	150–180	11–32
	glassine	6	25	150–180	2–3
Hot-melt wax	glassine	8	25	200–350	3–6
Cellulose nitrate	kraft	8	32	200–300	19–29
	glassine	6	22	200–300	6–13
Vinylidene chloride	kraft	6	25	250–300	5–8
	glassine	4	25	250–300	5–8
Polyethylene					
low density 1 mil	kraft	15	25	200–300	19–26
med density 1 mil	kraft	15	25	250–350	16–22
high density 1 mil	kraft	15	25	275–375	10–16

* Based on "Modern Packaging Encyclopedia," Modern Packaging, New York, p. 122 (1967).
† Water vapor transmission rate.

A special coating has been developed under Navy contract, Bureau of Naval Weapons Contract No. NOw 63-0155-C by Vincent Meier of Port Washington, N. Y. The contractor developed a method for use of materials with a desiccant-containing coating for military packaging under Method II of MIL-P-116. One approach suggested involves the use of packages made from paper impregnated with desiccating salts, then coated with a barrier material. With barrier outside and desiccating paper inside, the package interior stays dry. Meier calculates a cost saving of 40% over current Method II prodecures.

Lamination

Lamination processes involve the use of some kind of adhesive agent to provide a bond between two layers of material, each of which is in normal solid form. The laminating agents are in the standard form for such materials. They may be in solution or emulsion in water, or in solution in other solvents. Other materials can be heat-lique-fied. Most common laminations with paper involve the application of paper to aluminum foil. Usually, but not always, a third material is coated on one side or the other of the combination to provide heat-sealing and some additional barrier capability to the combination. Paper, laminating agent, and coating all help to enhance the barrier properties of aluminum foil in the finished package. It is a failing of aluminum foil that it does tend to thin, crack, and pinhole when it is stretched or flexed. The materials combined with it provide cushioning and bulk during flexure and toughness during stretch, which are of definite assistance in maintaining barrier properties through fabrication and handling.

The usual reason for laminating materials to paper is to obtain improvements in barrier properties, strength, or special surface decorative effects. A heat-seal coating must often be added to a lamination that fulfills the other functions. Laminations are constructed of multiple plies of sheet, and so paper-paper laminations are a logical expectation. Commonly, glassine/wax/glassine is used as a liner for dry cereal packages, among other things. This combination provides a paper carrier for the water vapor barrier wax. The wax is covered with a second layer of glassine so that the whole structure provides protection from water vapor (by wax) and from oxygen (by glassine) and is of course a satisfactory sealant for closure of the package. Laminating materials include many grades of paper, especially sulfite, glassine, and tissue; also aluminum foil, usually 0.00035 in. thick. Normally these are the only laminants to be found with paper. In some applications, notably in compliance with military packaging requirements, cellulose acetate or polyester (Mylar, Scotchpak, Celanar) films may be found laminated to paper. In certain applications for heavy-duty shipping containers, the lamination may involve paper-to-paper with asphalt as the laminating agent. Some typical laminations are listed in Table 13-3.

Laminating agents are the adhesives used to join two materials. The adhesives used run the full range from water-soluble dextrins to extruded resins or natural rubber. Paper-foil laminations are commonly joined with wax or polyethylene, and as mentioned previously the glassine/wax/glassine is widely used. One operating rule is that the agent must be able to be solidified easily. Thus, paper laminations can be made by most of the agents because the porous paper provides

TABLE 13-3

Some Typical Paper Laminations

1. a) 0.00035″ foil/17-lb microcrystalline wax/20-lb sulfite paper
    b) 0.00035″ foil/EVA-paraffin hot melt/paper
2. a) 25-lb pouch paper/0.5 mil polyethylene/0.00035″ foil/1.0 mil polyethylene
    b) 25-lb glassine/0.5 mil polyethylene/0.00035″ foil/1.0 mil polyethylene
3. a) 0.00035″ foil/2-lb casein adhesive/backing paper
    b) 0.0035″ foil/7-lb polyethylene/backing paper
4. Glassine/wax/glassine
5. Lacquer coated foil/adhesive/wet strength paper.

Note: It is conventional to write laminations from left to right starting with the outside. Thus lamination 1 has aluminum foil outside and paper nearest the product. The weights indicated are lb per ream of 300 ft² area. The (a) and (b) alternatives represent different ways of achieving approximately the same results.

ample opportunity for drying of solvents or extrusion of melts through the pores and consequent rapid cooling. A standard classification of laminating agents is

Aqueous Solution—dextrins, animal glue
Emulsion—rubber (latex), resinous mixtures
Solvent solution—resins or gums, synthetic and natural rubber
Heat-liquefied—waxes, wax blends, extruded resins such as polyethylene or
      nylone, asphalts.

Barrier Behavior

When we speak of the barrier behavior of coated or laminated papers, we tend to forget the hygroscopic nature of the paper substrate and the importance of that hygroscopicity to barrier performance. Considering this hygroscopicity and the moisture equilibrium curve for papers, which follows the nonlinear function of Fig. 13-1, it is to be expected that the water vapor and gas permeation values quoted at one relative humidity are not valid at another—even at the same temperature. Because water is absorbed by the cellulose and

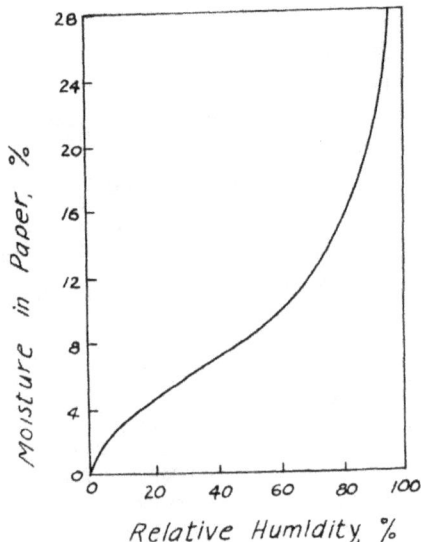

Fig. 13-1. Relationship of moisture content of paper to relative humidity [from W. Wink, *Tappi* **44** (6), 171A (1961)].

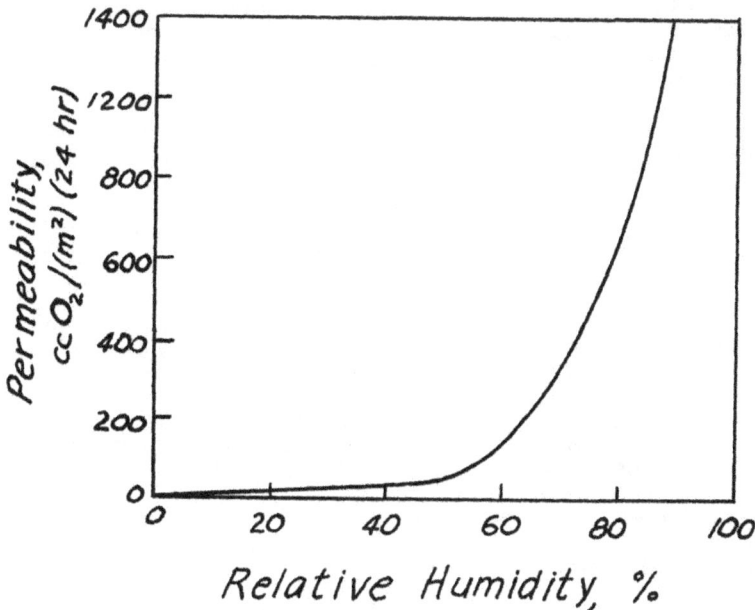

Fig. 13-2. Effect of relative humidity on permeability of polyethylene-coated parchment to oxygen at 73°F.

is a plasticizer for it, some fine barrier characteristics can be destroyed by increases in relative humidity. A polyethylene-coated parchment, when tested for oxygen transmission at various relative humidities at 73°F gave results displayed graphically in Fig. 13-2.

When dry, and even up to 50% RH, this material compares favorably with saran, nylon and polyester. At higher relative humidities, however, its barrier properties deteriorate rapidly.

PUBLICATIONS AND ORGANIZATIONS

One of the more elusive kinds of information in any field is the identification of useful and effective associations, organizations, and publications. It is probably impossible to be exhaustive in providing such information. However, a few such are outstanding and must be considered by anyone interested in packaging.

Publications

The following publications have proved to be of lasting value in keeping up with current developments in packaging:

Periodical	Coverage
Food and Drug Packaging 777 Third Avenue New York, N.Y. 10017.	General news with emphasis on food and drugs. Biweekly.
Materials Research and Standards 1916 Race Street Philadelphia, Pa. 19103.	Technical information on materials and testing. Monthly.
Modern Packaging 1301 Avenue of the Americas New York, N.Y. 10019.	General news with a good technical section in each issue. Monthly.
Modern Plastics 1301 Avenue of the Americas New York, N.Y. 10019.	General news with a great deal of technical-engineering information. Monthly.
Package Engineering 2 North Riverside Plaza Chicago, Ill. 60606.	Technical-engineering with much practical how-to-do-it-in-the-factory information. Monthly.
Packaging Design 1028 Connecticut Ave. N.W. Washington, D.C. 20036.	Surface and esthetic structural design. Bimonthly.
Paperboard Packaging 228 N. La Salle St. Chicago, Ill. 60601.	Technical-engineering with emphasis on paperboard. Monthly.
Paper, Film and Foil Converter 200 S. Prospect Ave. Park Ridge, Ill. 60068.	Technical-engineering with emphasis on the converter's point of view. Monthly.
Tappi 360 Lexington Ave. New York N.Y. 10017.	Science-engineering with emphasis on general paper product research and development. Monthly.
Printers Ink 501 Madison Ave. New York N.Y. 10022.	A printers-advertisers magazine that devotes considerable attention to the packaging-advertising interface. Bimonthly.
Abstract Bulletin of the Institute of Paper Chemistry Appleton, Wisconsin.	A good abstract but somewhat different coverage than *Packaging Abstracts*. Monthly.
Packaging Abstracts Published by Printing, Packaging and Allied Trades Research Association Randalls Road, Leatherhead, Surrey, England. (Available in U.S. from McGregor Magazine Agency, Mt. Morris, Ill. 61054.)	A fine abstracting service that reviews packaging and related publications from all over the world, including Eastern Europe. Monthly.

Not mentioned in the list are *Modern Packaging Encyclopedia* and *Modern Plastics Encyclopedia*. Both are 13th issues of their respective magazines. *Modern Packaging Encyclopedia* provides excellent information to the beginner in packaging. In fact, it serves as a text in a number of schools that teach packaging as a formal discipline. *Modern Plastics Encyclopedia* provides a wealth of technical-engineering data about plastics. It is an excellent companion to *Modern Packaging Encyclopedia* for those who wish to understand just a little better how plastics work. Some knowledge of organic chemistry is necessary to get full value from this book.

Organizations

A number of organizations serve packaging-oriented people and companies. Most of them deal at least indirectly with some application of paper.

American Ordnance Association 17th and HSt., N.W. Washington, D.C. 20006	Fulfills a liason function between the military and suppliers of protective packaging to the military.

American Society for Testing and
Materials
1916 Race St.
Philadelphia, Pa. 19103
    Committee D-6 on Paper and
      Paper Products
    Committee D-10 on Packaging
	Prepares industry standard tests and specifications based on current best practice in the industry. Committees D-6 and D-10 are responsible for national standards and for liasons with other organizations such as TAPPI.

Institute of Paper Chemistry 1043 E.S. River St. Appleton, Wisc. 54911	Industry-supported research and training organization.
National Flexible Packaging Assn. 11750 Shaker Blvd. Cleveland, Ohio	An association of converters, quite active in maintaining currency in the industry.
National Inst; tute of Packaging Handling and Logistic Engineers P.O. Box 7393 Washington, D.C. 20044	An association of suppliers to the military that maintains contact with the military on an informal basis.
Package Designers Council 299 Madison Ave. New York, N.Y. 10017	An association of independent package designers.
Packaging Institute 342 Madison Ave. New York, N.Y. 10017	A national association of companies from nearly all parts of the economy. Individuals are sought as professional members. Technical committees address themselves to solution of current problems.

Society of Packaging and
Handling Engineers
14 E. Jackson Blvd
Chicago, Ill. 60604

An association of regional chapters
that seeks to promote the individual
welfare of members.

Technical Association of the Pulp
and Paper Industry
360 Lexington Avenue
New York, N.Y. 10017

A scientific-technical-engineering association devoted to promotion of
technical advancement of the industry.
Work carried on in committees that
write test methods and specifications.

Numerous governmental organizations and agencies have influence
on decisions and activities in packaging. Among them are

Regulatory

Interstate Commerce Commission
Washington, D.C. 20423
Truck and Rail Transport

U.S. Coast Guard
Washington, D.C. 20226
Waterway Freight

Bureau of International Commerce
U.S. Dept. of Commerce
Washington, D.C. 20226
Export

Bureau of Customs
Treasury Department
Washington, D.C. 20226
Import

Federal Aviation Agency
Washington, D.C. 20553
Air Freight

Federal Aviation Agency and
International Air Transport Assn.
1060 University St.
Montreal 3, Quebec, Canada
Air Express

Post Office Department
Washington, D.C. 20260
Parcel Post and Air Mail

Food and Drug Administration
Department of Health, Education and Welfare

Foods, drugs, food additives including potential migrants from package, and parts of Fair Packaging Act.

Federal Trade Commission
Deceptive packaging and labeling including parts of Fair Packaging Act

Department of Agriculture
Consumer and Marketing Service
Meats and poultry

Agriculture Research Service
Pesticides and biologies for animals

Nonregulatory

Forest Products Laboratory
Madison, Wisconsin
A research organization devoted to improvement of wood and paper products.

TESTS AND SPECIFICATIONS

Tests and specifications are fundamental requirements for effective use of protective wraps, or any other package form. Many companies and many associations have codified tests and specifications by which they buy and sell. Excluding government, three organizations promulgate tests and specifications on a national level. These are The American Society for Testing and Materials, Packaging Institute, and The Technical Association of the Pulp and Paper Industry (TAPPI). These three organizations coordinate their efforts so that tests and specifications for the same properties or materials are uniform. For the most part, individual companies and associations use these tests or similar ones.

Government Specifications are readily available and are quite detailed as to requirements. Where a test method, or specification has been written by ASTM or TAPPI to cover a government requirement, this is normally quoted as being the approved method. The government specifications that apply to functional and protective papers are:

Greaseproofed, flexible (waterproof)	MIL-B-121
Water-vaporproof, flexible	MIL-B-131
Waterproofed, flexible, all-temperature	MIL-B-13239
Wrapping, laminated and creped	MIL-P-130
Wrapping, waterproofed, kraft	UU-P-271

Copies of these can be obtained from Supt. of Documents, U.S. Government Printing Office, Washington, D.C. 20402. The items designated MIL are Military Specifications, the one designated UU is a Federal Specification. The latter designation applies to specifications for items to be used primarily in nonmilitary applications.

chapter 14

Metal Foil Papers

JOHN H. DAVIES

ALUMINUM FOIL PAPERS

Aluminum-foil papers are different in many respects from other coated papers; the "coating" is actually a solid sheet of aluminum foil. Aluminum foil is manufactured by rolling operations in much the same way as other thin gauge metals. A billet or slab of aluminum is passed through breakdown mills and then transferred to finishing mills set to produce the desired gauge. In thicknesses greater than about 0.006 in. aluminum is known as strip or sheet, but when rolled below this thickness it is known as foil. Other materials, such as acetate films, cellophane, and vacuumized films and papers have incorrectly, at times, also been called foil, but these films should not be confused with genuine aluminum foil, which is an actual metal foil.

The aluminum that is used for rolling into foil is of approximately 99.5% purity, the remainder being principally iron and silicon. Usual standard gauges are 0.00025, 0.00035, 0.0004, 0.00045, 0.00065, 0.0007, 0.0008, 0.001, 0.0015, 0.002, 0.0025, 0.003, 0.004, and 0.005 in. Each of these gauges has a calculated square-inch yield per pound with a rolling tolerance of plus or minus 10%.

The standard width for the "Fancy Paper" industry is 26 in. Other widths, however, both wider and narrower for other uses are being produced and used. Since 1955 rolling mills for foils have been increased in width to allow a single width up to 72 in. to be produced. Lamination, however, of the foil to a substrate or backing paper or board is still limited to 52 to 60 in. because of the width limitations of present laminating equipment.

The gauge of foil to be used in manufacturing the metal-covered paper depends upon the final consumer use and the requirements of the completed product. Both light and heavy gauges can be readily laminated to all weights of paper from boards to tissue. The paper should not contain any ingredients that will reject the acceptance of the adhesive. It should be free from impurities and foreign matter that will result in blemishes on the surface of the foil-paper. The surface of the paper on the side to which the foil is mounted determines, to a great extent, the finished appearance. This is particularly true of the lighter gauges of foil, which are so thin and pliable that they tend to conform exactly to the contours of a rough paper surface, producing a surface on the foil that diffuses light, instead of giving a mirrortype reflection. Commercial grades with a calendered, machine-glazed or clay-coated finish usually have an appearance between these two extremes.

The foil itself as delivered from the rolling mill either has an absolutely smooth and glass-like finish known as a "bright," or a diffusing surface, known as satin, matte or dull. The bright surfaces result from contact with the highly polished steel rolls of the mill. The matte surface is obtained by rolling two sheets together, back to back. Selection of either the bright or satin foil surface depends solely on the finished effect desired. With heavier weights of foil, such as 0.002 and 0.004 in. the surface of the paper does not play such an important part, because the foil tends to bridge over the depressions and thus compensates somewhat for roughness of the paper surface.

MOUNTED FOIL

The finished appearance of the mounted foil is also affected by the adhesive used in gluing the foil to the paper. One of the most widely used types of laminating adhesives today is based on casein-latex. Other types of adhesives, such as sodium silicate (water glass), wax and resins, are being used. The selection of an adhesive is dependent upon the end use of the finished lamination. A sodium silicate is used when an inexpensive lamination is required, and when moisture or water-resistance is not a factor. A casein latex type is used when more water-resistance is required. Wax, as an adhesive, is used in laminating lightweight papers and foil for use as candy wraps in protecting the product from moisture. Each of the adhesives has its place in producing a finished product for a particular need or specification.

Several methods of bringing together the web of metal foil and the web of paper to form the metal-paper composite are in use. The most common method is carried out in the tunnel-type laminator, which consists of a mounting head, a heated and ventilated tunnel approximately 35 ft in length, and a rewind unit. The webs of aluminum foil and paper are brought together at the head of the machine, the adhesive being applied by pickup and applicator rollers from the glue pan directly to either the web of metal or the paper. Whether the adhesive should be applied to the foil or to the paper depends upon the type of paper, foil, adhesive and the type of laminator being used. After the webs are brought together between pressure rolls, they travel through the tunnel or over steam-heated drier drums to drive off the moisture contained in the adhesive so that the moisture will not be trapped within the roll of the finished laminated product.

Good automatic control of this operation is important, since the incorrect moisture content may cause curling. Excess moisture in the finished lamination may also cause corrosion of the surface of the foil. This corrosion usually takes the form of minute, irregular white spots, which appear much as if little worms had eaten out the surface of the foil. It is also important that after the metal paper has been put into stock, it is kept in a dry and ventilated place and handled in the same manner as fine printing paper, so that the paper will not reabsorb any moisture from the air. If this does occur, corrosion may appear within the roll because the moisture picked up will remain in the paper unless again driven off by heat. Curling is due to the fact that the paper backing expands and contracts with changes in moisture content, whereas the metal foil remains dimensionally stable and cannot pick up or lose moisture. The combined metal paper curls in the same way as fine coated cover or printing papers and should be handled as such.

COLORING OF FOIL

Coloring of the foil can be accomplished by staining or lacquering or by dyeing the surface after an oxidizing and etching process, such as an anodic treatment. Anodic treatment yields brilliant colors, but is not in commercial use in this country to color aluminum foil in roll form. The usual coloring operation is similar in principle to a coating or over-all printing operation in that the color is picked up from the reservoir by means of rollers and applied uniformly to the foil surface. The foil then travels through a ventilated tunnel over a heated bed to drive off the solvents. The lacquer may be applied either before, after, or during the laminating process. A single-color lacquer is usually applied by a gravure cylinder in the laminating process, or after laminating in a special gravure press, when more than one color is desired.

Numerous synthetic lacquer bases can be used; however, a vinyl or nitrocellulose is used as the principal base for foil coloring. The colors or color combinations are unlimited, as is the number of possible formulations of bases, resins, solvent, and plasticizers that can be used to obtain the desired decorative effect. Clear lacquers are used to facilitate printing and to minimize finger marking. In all these lacquering operations, the foil acts as a complete barrier to the solvent so that all evaporation must take place from the surface. In view of this situation and because of the short time available in the tunnel for drying, considerable skill is required in blending the solvent to avoid blushing, bubbles, and incomplete coverage.

Color shades may be commercially matched from samples of other foils, papers or cloth. There is, however, a difference in dyestuffs from various manufacturers and, therefore, it is not always possible to obtain an exact match in tone or depth, even though the sample being matched is a foil sample. Many variables are involved in color matching: the brightness of the foil surface must be equal to that of the sample being matched, the sample to be matched might be old and somewhat faded and, as mentioned, the dyestuffs might be different. The initial matching of a color is usually time-consuming and expensive because of the scrap produced before the proper color density, hue, tone, and shade are obtained. At times it may not be possible to obtain an exact match, and laboratory samples may have to be submitted prior to actually running on production equipment.

EMBOSSING FOIL

Laminated foil and paper can be readily embossed to produce attractive decorative patterns. Embossing designs are limited only by quantity and variety of the embossing rolls available. Embossing is done in the same way as on other types of coated and uncoated papers. From the selected design the engravers engrave a blank steel roll to the specification of the embossing machine. The expense involved for a special design on a small order is prohibitive, but for an embossing design already available the cost is reasonable. Because of the high cost of special rolls and of the engraving operation, usually only one of the two embossing rolls is engraved, the second being a special paper roll, having exactly twice the diameter of the steel roll. This paper roll is operated wet against the engraved roll, under high pressure, until its surface becomes an exact negative of the embossed design. If desirable, print attachments can be used so that the foil can be print-embossed or topped during the embossing operation.

In selecting a design for embossing, attention should be paid to obtaining the greatest number of highlights or reflecting areas on the embossed foil to give the finished product sparkle and life. Trade marks or "name" embossings may be used in the same manner as described. For certain purposes a "rough" or bark-cloth type of design may be used. Many orders for paper and foils have been lost or obtained by the purchaser liking or disliking a particular design or pattern for his end use. There is no definite rule as to what a certain class of trade may like or dislike in embossing designs as this matter concerns individual taste and ultimate use.

The basic ideas of embossing, more or less, hold true for printing of foils. It should be kept in mind that in either embossing or printing designs, the beauty of foil rests in its clean, sparkling, bright finish and that an embossing that tends to dull the surface or an over-all opaque printing that hides much of the surface of the foil detracts from its visual appeal. Printing on foil has had a tremendous appeal to the buying public; great advances have been made in this field in the last few years. High-speed rotogravure presses now print many types of labels, magazine inserts, gift wrappings, advertising pieces and food wraps and bags. The foil surface of the lamination is usually coated with an extremely light wash consisting of a base carrier and shellac known in the trade as a "wash" or "print-treat" coat. This coating permits the printer to set his equipment at maximum speed as the coating acts as a base for adherence of his inks to the foil surface.

PRINTED FOIL

Foil laminated to boards in thickness of 0.008 in. and more is now being printed by the offset lithographic process with considerable success. For this method of printing the foil surface is also coated but the coating is a much heavier coat than that used for the rotogravure method. The coating generally consists of a vinyl lacquer, either clear or colored with some exceptions, however, where the printer specifies a nitrocellulose-type lacquer. This lacquer serves the same purpose as the "print-treat" coat for rotogravure in that it serves as a base for adherence of inks, aids in preventing finger marks and (when a colored lacquer is used) becomes part of the printing layout and decorative effect.

The inks used for either method should be those specifically made for the coating used on the foil and the method of printing. A good ink house should be consulted so that experts can determine the best ink for a particular type of press, coating, and method of printing.

GAUGING OF FOIL

The most generally used foil in laminations is in the range of gauges 0.00025 to 0.00035 in. These gauges are the most economical to use because of the cost of the foil in relation to the yield. Each gauge of foil has a calculated square inch yield per pound, which is based on the specific gravity of aluminum (see Table 14-1). The most economical lamination of these gauges is to papers within the range of 20 to 45 lb, bleached white sulphates, natural brown sulphate, or groundwood.

TABLE 14-1
Area Yield Figures
Officially Adopted by The Aluminum Association

Thickness, in.	Covering area, in.^{2}/lb	Thickness, in.	Covering area, in.^{2}/lb
0.00017	60,300	0.0008	12,800
0.0002	51,300	0.00085	12,100
0.00025	41,000	0.0009	11,400
0.0003	34,200	0.00095	10,800
0.00035	29,300	0.001	10,250
0.0004	25,600	0.0015	6,830
0.00045	22,800	0.002	5,130
0.0005	20,500	0.0025	4,100
0.00055	18,600	0.003	3,420
0.0006	17,100	0.0035	2,930
0.00065	15,800	0.004	2,560
0.0007	14,600	0.0045	2,280
0.00075	13,700	0.005	2,050
		0.0055	1,860

The covering areas shown in this table are based on a specific gravity of 2.70 or a density of 0.0976 lb/in.^{3}—which is equivalent to 10.25 in.^{3}/lb. Thus, with the foil thickness known, covering area per pound of foil is determined by the formula:

1. *To convert thickness into area*

$$\frac{10.25 \text{ in.}^3/lb}{\text{in. thick}} = \text{in.}^2/lb$$

Example: $\frac{10.25}{0.00035} = 29,285$—rounded to 29,300

2. *To convert area into thickness*

$$\frac{10.25 \text{ in.}^3/lb}{\text{in.}^2/lb} = \text{in. thick}$$

Example: $\frac{10.25}{29,300} = 0.00035$

When a paper weight is given in a lamination it refers to the basic weight of the paper prior to lamination, and as manufactured and sold by the paper mill. Therefore a 30-lb paper has a basis weight of 30 lb, $24 \times 36 \times 500$, or as usually written, 24×36—30/500. This simply means that 500 sheets of the paper 24 by 36 in. have a weight of 30 lb and is the basis mill ream weight. Only a simple calculation is required to convert the weight of the paper from this ream basis to any other ream basis. The "Fancy Paper" trade usually employs a ream based on 500 sheets size 20 by 26 in. ($20 \times 26 \times 500$). Converting the basic weight of 30-lb to this ream size the backing paper of the lamination will have a weight of 18 lb; it is, however, still a basic 30-lb paper ($24 \times 36 \times 500$).

All paper and board run on high-speed laminators are laminated in roll form and therefore the grain direction of the finished lamination is determined by the paper the grain of which is always parallel to the length of the paper. If a ream is expressed as being $20 \times 26 \times 500$ in 26 in. wide rolls, the grain of the lamination would be parallel to the 20-in. dimension. In any given sheet size, if the actual width is known the grain is always the opposite running parallel with the length.

The usual width of metal paper in the "Fancy Paper" industry is 26 in. The usual widest continuous web of metal that is laminated is 52 in. As mentioned previously, the foil mills are rolling foil to 72 in. however, few laminators are now laminating any wider than this because of the limiting width of their existing laminators. Narrower widths, of course, may be supplied and the wide stock slit or sheeted to smaller sizes as required.

FOIL LABELS

The uses for metal paper are many and cover a wide range of industries, from candy to electronics, from home insulation to gift-wrap and greeting cards, picture frames to food packaging, background displays and candyboxes to beer labels and oil to fruit juice canisters. Several of these products are discussed in some detail in the following paragraphs.

Bottle Labels

Beer and bottle labels printed by the rotogravure method: The foil gauge is usually 0.0003 in. glue-laminated to a 33.2-lb (basis $25 \times 38 \times 500$) groundwood paper with a print-treat coating on the foil surface. It is usually supplied to the printer in large-diameter rolls in the width best for the size of the label to be printed to give him the most economical printing width. It is run at high speed, printing four or more colors, sheeted at the end of the press and then die cut to the shape and size required. Moisture control during lamination is desirable to prevent curling of the sheets prior to die cutting, and in the stacks of the die cut labels.

Neck labels

The 0.00035-in. foil laminated to a 20-lb paper is being used as a bottleneck foil. It is used in silver or colored finish, embossed, and cut in rectangular shape or die-cut to a specific shape. This is a small band that goes around the neck of the bottle and is used by the brewers and also by the soft-drink bottlers. The embossed pattern is usually a stock design or can be a brand name embossing by arrangement with the mill producing the material.

Special Label Stocks

These may be made by using the 0.00035 in. either glue-or wax-mounted to litho coated papers. The paper side can be coated with a pressure-sensitive coating or a heat-seal coating. When combined with glassine papers and either heat-seal-coated or extruded with a light film of polyethylene, the material can be fabricated into heat-sealable bags for potato chips, popcorn, dehydrated soup mixes, and powders.

GIFT WRAPS

The foil gauge generally used for gift wraps is 0.00025 in. solid glue-laminated to several weights of paper; those mostly used are 20-lb, 25-lb and 30-lb, basis $24 \times 36 \times 500$. The adhesive used is a silicate (water glass) and the paper is either a bleached white sulfate or sulfite. It may be supplied to gift wrap manufacturers in several ways, large mill rolls in silver treated for printing, embossed in silver or colors, or in printed designs. The gift wrap manufacturer will, in the case of print treated stock, print this in his designs, use this and the colored embossed stock to rewind in resale rolls of desired lengths.

BOX COVERINGS

Boxmakers also use metal paper for their box coverings. The usual specification for this purpose is 0.00035 gauge, laminated to a 30-lb paper. The manufacturer who uses the boxes to package his product usually selects the design or the color to be used for the boxes. Here, again, the color and embossing design play an important part in the actual sale of the metal paper and are usually selected to conform to the item to be packaged. When the design and color are selected, the boxmaker will then purchase his stock for covering the box. Many types and kinds of boxes have been made with metal paper cover. In some cases, the complete box is covered with one paper and in others only the top or bottom section is covered with one color metal paper and the other part coated with a contrasting metal paper. The finish may be plain, embossed or printed. The foil may be printed on high-speed rotary presses, by the silk-screen method, or by regular flat-bed press. Some very beautiful effects have been obtained by both embossing and printing on the foil in multiple colors. Here, again, embossing design and printing designs are limited only to the embossing and printing rolls available at any one of the mills. Metal-paper-covered boxes have been made for such products as jewelry, food, candy, cosmetics, liquor, and writing paper. In fact, almost any manufacturer that uses a decorative box in packaging his product is a potential user of metal paper as a box covering.

SPIRAL-WOUND CONTAINERS

Aluminum foil laminated to several weights of natural northern kraft paper serve the canister manufacturer of spiral wound containers in two ways. When the foil surface is treated with a "slip-treat" coating, it is used as the inner ply of the canister. The "slip-treat" coating allows the foil to be wound on the mandrel at high speeds without binding or generating excessive heat because of friction as the canister is being formed. The foil now becomes the inner lining of the tube or canister and helps to contain the product to be packaged.

When the foil surface is treated with a "print-treat" coating, it is usually sold and shipped to a rotogravure printer. It is printed on high speed rotogravure presses and then sent to the canister manufacturer, who uses the stock as the outer ply, which then becomes the printed label of the finished canister. The printed label serves as a decorative printed label and also as an aid in protecting the product to be packaged. The foil gauge can be either 0.0003 or 0.00035, the paper for the liner 25-lb; the paper for the label can be 30-lb or 40-lb. The different combinations of foil and paper depend upon a particular manufacturer and the product that is packaged. Many products are packaged in this type of a container. Some of these are frozen fruit juice, refrigerated biscuits, oil, grease, powders, and insecticides.

GREETING CARD STOCK

A steadily increasing amount of aluminum foil solidly glue-laminated to solid bleached white sulfate board is being used in the greeting card industry for greeting cards of every nature; the foil gauge is usually 0.00035 or 0.00030 and laminated to a 0.010-point board with a casein latex-type adhesive. The usual finish is either silver or gold, and supplied in the bright or dull (matte) finish. The card is usually printed or embossed on the foil surface and the greeting is printed on the reverse or unlaminated side of the board.

The type of lacquer used for the silver (clear) or gold coating is very important as most cards are printed by the offset method. Some offset printers prefer the nitrocellulose base; others, a vinyl base. The printer must state his preference in lacquer prior to coating as an incorrect base can cause many difficulties. The stock is supplied in large sheets to be fed into the large size offset presses, and must be trimmed and squared.

INSULATION

Foil gauges between 0.00025 and 0.00035 in., laminated with a silicate-type adhesive to natural brown kraft papers in weights between 25 to 40 lb (basis $24 \times 36 \times 500$) are used by the insulation manufacturers to form the facing of batts that are filled with either blown mineral wool or spun glass wool. The foil surface acts as a heat reflector and a moisture barrier. When the foil surface is perforated it allows the passage of air, yet retains its reflective qualities. In the batt form it is commonly used for insulation of walls and ceilings of homes. It is being combined with building and insulation boards to act in conjunction with these boards for insulation purposes.

CONFECTIONERY FOIL

The 0.00035-in. foil, when combined with a lightweight wax paper, is used extensively in the confectionery field. It is used for the packaging of chocolate bars and patties. A glassine paper may also be used as a backing. The use of the wax or glassine paper with the foil is to aid stripping the wrapper from the chocolate since most other types of paper have a tendency to adhere to the chocolate surface. Hard candies are being wrapped in this lightweight waxed paper and the foil surface is printed with a decorative design or with a brand name.

For many years chewing gum has been packaged in a combination of foil and waxed paper.

APPLICATIONS

The applications of the foil and paper combinations are so varied that space here does not permit the discussion of the combinations or uses. Needless to say, specifications for the combination to be supplied are extremely important, both to the mill manufacturing the item and the customer. Many questions should be definitely answered when a specification is being discussed. All are important.

1. What is the end use?
2. What is expected of the metal paper combination?
3. Is the foil surface to be printed? If so, how—gravure, offset, letter press?
4. Should a vinyl coating or a nitrocellulose be supplied?
5. Is there a general low-odor requirement involved?
6. Is an FDA clearance for the lamination and the coating involved?
7. Is a fade-resistant requirement involved? How many fade hours are expected?
8. Has the paper mill basis weight of the paper and board been discussed and understood?
9. Has the adhesive for the particular combination and end use been clarified?
10. Have proper packing specifications of either roll or sheeted stock been determined?

Both the mill and the customer should submit samples for trial if there is anything off beat in the way of specifications. This procedure will allow both the mill and customer to determine whether the combination is satisfactory prior to expensive mill production.

The field of uses for foil without paper backing is quite extensive. The foil can vary in thickness between 0.00025 and 0.00600 gauge. The very light foils are used in electrotytic condensers. Heavier weights, such as 0.00070, may be used for protective purposes in the home by the housewife in various ways and are handled in the same way she uses wax papers. Aluminum foil utilized in this manner will help to preserve foods for a longer period of time, than when those foods are stored in the conventionsl manner. It may also be used in broiling, roasting, and baking.

Foil in intermediate gauges unbacked or backed with a cellophane film is used in the florist trade and is known as florist foil. It is also used as "Tree Foil" for the manufacture of foil Christmas trees. In hard temper and in many colors, it is "chopped" in small squares or as fine as powder to be used in vinyl coatings, floor tiles, and spray paints; in this form it is known as "flocking foil" or "glitter".

On the basis of the specific gravity of aluminum, the aluminum foil has a calculated square-inch yield per pound with a rolling tolerance of plus or minus 10%. The various gauges mentioned previously with their approximate square-inch yield per pound are listed in Table 14-1.

Index

www.ingramcontent.com/pod-product-compliance
Lightning Source LLC
Chambersburg PA
CBHW021026210326
41598CB00016B/926